Y0-BYB-355

IRPA GUIDELINES ON PROTECTION AGAINST NON-IONIZING RADIATION

The Collected Publications of
The IRPA* Non-Ionizing Radiation Committee

*The International Radiation Protection Association

Edited by
A.S. Duchêne
J.R.A. Lakey
M.H. Repacholi

PERGAMON PRESS

Member of Maxwell Macmillan Pergamon Publishing Corporation
New York • Oxford • Beijing • Frankfurt
São Paulo • Sydney • Tokyo • Toronto

Pergamon Press Offices:

U.S.A.	Pergamon Press, Inc., Maxwell House, Fairview Park, Elmsford, New York 10523, U.S.A.
U.K.	Pergamon Press plc, Headington Hill Hall, Oxford OX3 0BW, England
PEOPLE'S REPUBLIC OF CHINA	Pergamon Press, Xizhimenwai Dajie, Beijing Exhibition Centre, Beijing 100044, People's Republic of China
GERMANY	Pergamon Press GmbH, Hammerweg 6, D-6242 Kronberg, Germany
BRAZIL	Pergamon Editora Ltda, Rua Eça de Queiros, 346, CEP 04011, Paraiso, São Paulo, Brazil
AUSTRALIA	Pergamon Press Australia Pty Ltd., P.O. Box 544, Potts Point, NSW 2011, Australia
JAPAN	Pergamon Press, 8th Floor, Matsuoka Central Building, 1-7-1 Nishishinjuku, Shinjuku-ku, Tokyo 160, Japan
CANADA	Pergamon Press Canada Ltd., Suite 271, 253 College Street, Toronto, Ontario M5T 1R5, Canada

---

Library of Congress Cataloging-in-Publication Data

The IRPA guidelines on protection against non-ionizing radiation : the collected publications of the IRPA Non-ionizing Radiation Committee / editors, A.S. Duchêne, J.R.A. Lakey, M.H. Repacholi.
p. cm.
Includes index.
ISBN 0-08-036097-1 (hardcover)
1. Nonionizing radiation--Safety measures. I. Duchêne, A. S. II. Lakey, J. R. A. III. Repacholi, Michael H. IV. International Radiation Protection Association. International Non-Ionising Radiation Committee.
RA569.3.177 1990
363.17'9--dc20 90-7315
CIP REV

Printing: 1 2 3 4 5 6 7 8 9 Year: 1 2 3 4 5 6 7 8 9 0

Printed in the United States of America

The paper used in this publication meets the minimum requirements of American National Standard for Information Sciences -- Permanence of Paper for Printed Library Materials, ANSI Z39.48-1984

IRPA GUIDELINES ON PROTECTION AGAINST NON-IONIZING RADIATION

Pergamon Titles of Related Interest

Hall: RADIATION AND LIFE
Turner: ATOMS, RADIATION, AND RADIATION PROTECTION
Cember: AN INTRODUCTION TO HEALTH PHYSICS
Turner: PROBLEMS AND SOLUTIONS IN RADIATION PROTECTION

Related Journals

(Free sample copies available upon request.)

HEALTH PHYSICS
ANNALS OF THE ICRP
INTERNATIONAL JOURNAL OF RADIATION ONCOLOGY-BIOLOGY-PHYSICS
THE ANNALS OF OCCUPATIONAL HYGIENE

IRPA/INIRC MEMBERSHIP: 1977 TO 1988 AND 1988 TO 1992

During the preparation of the documents collected in this book, the following members served on the International Radiation Protection Association/International Non-Ionizing Radiation Committee (IRPA/INIRC) at different periods from 1977 to 1988:

H. P. Jammet, Chairman (France)
J. Bernhardt (Federal Republic of Germany)
B. F. M. Bosnjakovic (The Netherlands)
P. Czerski (Poland, then U.S.A.)
M. Faber (Denmark)
M. Grandolfo (Italy)
D. Harder (Federal Republic of Germany)
B. Knave (Sweden)
G. Kossoff (Australia)
J. Marshall (United Kingdom)
M. H. Repacholi (Canada, then Australia)
D. H. Sliney (U.S.A.)
J. A. J. Stolwijk (U.S.A.)
J. C. Villforth (U.S.A.)
G. M. Wilkening (U.S.A.)
A. S. Duchêne, Scientific Secretary (France)

For the period 1988 to 1992, the membership of the IRPA/INIRC is the following:

M. H. Repacholi, Chairman (Australia)
J. Bernhardt (Federal Republic of Germany)
B. F. M. Bosnjakovic (The Netherlands)
L. A. Court (France)
P. Czerski (U.S.A.)*
M. Grandolfo (Italy)
B. Knave (Sweden)
A. F. McKinlay (United Kingdom)
M. G. Shandala (U.S.S.R.)
D. H. Sliney (U.S.A.)
J. A. J. Stolwijk (U.S.A.)
M. A. Stuchly (Canada)
L. D. Szabo (Hungary)
H. P. Jammet, Chairman Emeritus (France)
A. S. Duchêne, Scientific Secretary (France)

---

* Professor Przemyslaw A. Czerski, charter member of the INIRC, died on April 15, 1990 in Silver Spring, Md. (U.S.A.). He was a pioneer investigator in the effects of non-ionizing radiation on biosystems and the assessment of the potential hazards associated with such exposure. As a fervent promoter of international cooperation, Prof. P. Czerski played an active part in the establishment of the INIRC and the development of its activities. His broad scientific knowledge and his tireless energy made him one of the main contributors to the present publication.

CONTENTS

FOREWORD

THE LOGO of the International Radiation Protection Association (IRPA) that appears on the cover of this book has a strong resemblance to the radiation warning symbol, the trefoil, which is widely used to warn the observer that a radiation hazard is present. No such warning is intended by the use of this logo that was designed to represent the primary purpose of the IRPA. The logo should be recognised as a rotated version of the warning symbol, emphasizing that IRPA's position is on the safe side of the radiation hazard.

This book is also representative of IRPA's primary purpose, which is to serve as a medium for international communication and cooperation in radiation protection with the goal of advancing sound and effective radiation protection in all parts of the world. The guidelines contained in this book are the culmination of over 12 years work of the International Non-Ionizing Radiation Committee (INIRC) under the leadership of the Chairman Emeritus, Professor H. Jammet, and the Chairman Dr. M. Repacholi, together with the Scientific Secretary Mrs. Annette Duchêne. As publications director until 1988 and now as president I have had the privilege of working with this team through publication of the Interim Guidelines in *Health Physics* up to this revised collection. Throughout this period, IRPA funds have supported the committee and have contributed to the cost of publications and administration of the committee.

On behalf of the IRPA I thank all who have contributed to these guidelines. We are grateful for the encouragement and support of the International Agencies and particularly the World Health Organization, the United Nations Environment Programme, the International Labour Office, and the Commission of the European Communities.

IRPA now has over 15,000 individual members who are represented by 31 associated societies in over 36 countries. In 1991 the association will attain its 30th year of service to the international community. It is governed by its members through a representative general assembly, which is the keystone of the International Congress of the IRPA held every 4 years.

Publications are an important part of this service and, in addition to the hospitality of the pages of the Journal of the Health Physics Society—*Health Physics*—in which these guidelines were first published, a bulletin is circulated quarterly to member societies and supporting agencies. The bulletin and *Health Physics,* which is available to IRPA members at a special rate, contain news and notices of interest to the profession. Members can gain access to other journals in the field of radiation protection at a reduced rate. In addition to the International Congress, IRPA also sponsors regional congresses organised by two or more member societies; there is also increasing interest in the provision of topical sessions, meetings, and workshops to assist in international cooperation and communication.

The dedicated work that was required in the writing of these guidelines was done by the members and scientific secretary of the INIRC. They will support me in expressing gratitude for the contributions of many members of the IRPA and the editorial staff of *Health Physics,* who have read and commented on these guidelines.

The work of the committee was conducted with the encouragement of the IRPA Executive Council and I pay special tribute to the officers: the presidents—C. Polvani, C. C. Palmiter, Z. M. Beekman, and M. W. Carter; the vice-presidents—G. Cowper, H. Jacobs, and C. B. Meinhold (responsible for appointments to the INIRC); the treasurers—P. Courvoisier, W. Hunzinger, and R. Maushart; the executive officers—J. Horan, G. Bresson, and Chr. Huyskens; and to my successor as publications director—J. E. Till; together with all of the executive council. We want to express our continued support for this important international venture and our wish that the committee will become independent, retaining a special relationship with IRPA and consolidating the international status that it has achieved.

All who have supported and participated in the work of the INIRC of IRPA share with me the hope that these pages will provide a lasting contribution to the advancement of sound and effective radiation protection in all parts of the world.

J. R. A. LAKEY
President of the International Radiation Protection Association
Meopham, England, December 1989

PREFACE

PROTECTION against radiation is a broad field including ionizing and non-ionizing radiation. At its International Congress of Radiology in 1928, the International Society of Radiology created two special commissions: the International Commission on Radiation Units and Measurements (ICRU) and the International Commission on Radiological Protection (ICRP). The ICRU covers the whole field of ionizing and non-ionizing radiation. The ICRP, however, refused to include protection against non-ionizing radiation (NIR) in its field of competence.

Over 15 years ago, during the Third International Congress of the International Radiation Protection Association (IRPA) in Washington in 1973, a session on protection against NIR was organized with H. Jammet as rapporteur. From 1973 to 1977, with the constant and effective influence and support of Carlo Polvani, President of IRPA, the Executive Council of IRPA decided first to establish a Working Group (1974), then a Study Group (1975–1976), to review the situation in the field of NIR.

During the Fourth International Congress of IRPA in Paris in 1977, the IRPA General Assembly decided to modify its constitution and to include explicitly NIR in its field of activities. Then, upon the IRPA Executive Council's proposal, the general assembly created the International Non-Ionizing Radiation Committee (INIRC). This committee is composed of a chairman, a scientific secretary and 12 members, selected on the basis of scientific qualification and expertise within appropriate disciplines.

To achieve the tasks that the IRPA had imparted to it, the INIRC agreed that the different types of NIR would be dealt with successively, each according to the following scheme:

1. Compile all available background information and determine the basic health criteria for the relevant NIR;
2. Develop guidelines on appropriate limits of exposure for workers and the general public;
3. Provide guidance on the practical measures to be taken for the safe use of the NIR considered.

During the 12 years that have elapsed since INIRC was created, its members devoted their efforts to carrying out this programme and for the time being more than 15 reports have been produced by IRPA/INIRC either alone or in cooperation with other international organizations.

To assess the health risk associated with any given NIR, it is first necessary to analyse all available data relating to its physical characteristics and biological effects, the sources in use, the resulting levels of exposure, and the people at risk. In cooperation with the United Nations Environment Programme and the World Health Organization, the IRPA/INIRC published six Environmental Health Criteria Documents:

- EHC 14: Ultraviolet Radiation (1979)
- EHC 16: Radiofrequency and Microwaves (1981)

- EHC 22: Ultrasound (1982)
- EHC 23: Lasers and Optical Radiation (1982)
- EHC 35: Extremely Low Frequency (ELF) Fields (1984)
- EHC 69: Magnetic Fields (1987)

The recommendation of appropriate exposure limits for workers and for members of the public is the chief objective that has been assigned to the INIRC by the Executive Council and the General Assembly of IRPA. The IRPA/INIRC Guidelines on limits of exposure to the different NIR are established on the basis of the scientific data collected for the relevant Environmental Health Criteria and on any later published research data. The purpose of the guidelines is to deal with the basic principles of protection against the relevant NIR, so that they may serve as guidance to the various international, regional, and national bodies as well as the individual experts who are responsible for the development of regulations, recommendations, or codes of practice to protect the workers and the general public. All the guidelines are published in *Health Physics.*

The present book contains the IRPA/INIRC Guidelines for:

- ultrasound;
- radiofrequency electromagnetic fields (100 kHz–300 GHz);
- ultraviolet radiation;
- laser radiation of wavelengths between 180 nm and 1 mm;
- 50/60 Hz electric and magnetic fields.

In addition, the book also contains the following documents produced by INIRC:

- review of concepts, quantities, units, and terminology for NIR protection;
- a statement on alleged radiation risks from visual display units;
- a statement on fluorescent lighting and malignant melanoma.

As chairman of IRPA/INIRC from 1977 to 1988, I want to sincerely thank all the members of the INIRC for their cooperative work in producing the documents, and especially the Scientific Secretary Mrs. A. Duchêne who was in charge of the final revision.

I am sure that under the chairmanship of the new chairman, M. H. Repacholi, the IRPA/INIRC will be able to continue and to develop its work to achieve the best protection against NIR for workers and members of the public.

I want also to thank the past and present presidents, vice-presidents, officers and members of the executive council of IRPA for their support of INIRC activities.

H. P. JAMMET
IRPA/INIRC Chairman Emeritus

CHAPTER 1

GUIDELINES ON PROTECTION STANDARDS FOR EXPOSURE TO NON-IONIZING RADIATION

THE INTEREST that the International Radiation Protection Association (IRPA) has shown concerning protection problems arising from the rapidly expanding use of non-ionizing radiation (NIR) stems from the internationally recognized need to arrive at an agreement on a realistic and effective general policy in this field. A comparison of presently existing national recommendations or codes of practice shows that wide variations exist in the basic approach to protection against the hazards resulting from NIR and therefore in the recommended standards. The authorities responsible for establishing safety regulations in countries where they do not yet exist find it difficult to decide on specific exposure limits, given the diversity of the basic policies and limits that have been adopted in countries having already promulgated such legislation and the technical complexity of certain solutions.

The IRPA took on the responsibility for activities in the field of NIR by forming a working group in 1974. Because of the importance of this problem, at the 1977 IRPA Congress in Paris, the IRPA created the International Non-Ionizing Radiation Committee (INIRC) with the following commitments:

- to assess present knowledge on the biological effects of NIR;
- to develop background documents and internationally acceptable recommendations;
- to explore with other international organizations and agencies the ways and means for furthering NIR protection activities.

The IRPA/INIRC, in cooperation with the Environmental Health Division of the World Health Organization (WHO), has undertaken responsibility for the development of health criteria documents on NIR. These form part of the WHO Environmental Health Criteria Programme, which is sponsored by the United Nations Environment Programme (UNEP). The documents include an overview of the physical characteristics, measurement and instrumentation, sources and applications of NIR, a thorough review of the scientific literature on biological effects, and evaluations of the health risks of human exposure to NIR.

Six reports dealing respectively with environmental health criteria relating to ultraviolet radiation (UN79), radiofrequency and microwaves (UN81), ultrasound (UN82a), lasers and optical radiation (UN82b), extremely low frequency (ELF) fields (UN84), and magnetic fields (UN87) have been published by WHO under the joint sponsorship of UNEP, WHO, and IRPA. Numerous experts from various countries and institutes have contributed to this project through their comments and suggestions before each final document was produced by a multidisciplinary group of experts representing a wide range of scientific opinions.

These criteria then become the scientific data base for the development of the IRPA/INIRC guidelines on limits of exposure to the different types of NIR that are presented in the following chapters.

The IRPA/INIRC furthermore cooperates with the International Labour Office (ILO) to provide practical guidance for the protection of workers against hazards due to NIR in the working environment (ILO 1985, ILO 1986).

The INIRC is funded by the IRPA. In addition, the support received from the ILO, the WHO, the UNEP, and the Commission of the European Communities is gratefully acknowledged.

INTRODUCTION

The present report will first give a review of some general concepts. It will then concentrate on defining the basic rationale underlying an internationally acceptable protection policy against NIR.

While ionizing electromagnetic radiation clearly refers to all radiation having individual photon energies sufficient to ionize atoms, an absolute statement regarding NIR is difficult. Normally, lower-energy photons (i.e., $\leq$12 eV) have insufficient energy to release a bound electron from an atom. However, in solid-state matter, electrons can be released by much lower photon energies. Furthermore, high-power laser beams can be focused to produce plasma-ionized matter. In general, the term NIR includes all electromagnetic radiations having a wavelength equal to or greater than 10^{-7} m, that is, ultraviolet (UV) radiation (100–400 nm), visible light (400–760 nm), infrared radiation (760 nm–1 mm), all radiofrequencies from the upper limit of the microwave spectrum (300 GHz or 1 mm) down to the longest radiowaves (100 kHz or 3 km), as well as the extremely low frequency range (below 300 Hz).

Such radiation can be emitted continuously or intermittently and the modulations that affect the frequency, amplitude, or pulse can modify some of the health aspects. Highly directional beam sources such as lasers may present special problems.

For protection purposes, the area of NIR is generally extended to electrostatic and magnetostatic fields. In addition, the IRPA/INIRC has included infrasonic and ultrasonic radiation in its activities because the problems related to protection of workers and the general public, which are similar to those of electromagnetic radiation, have not been specifically dealt with until now by any international organization.

BASIC CONCEPTS

In assessing health risks, the concepts of effect and damage are often confused and should be distinguished.

Whenever there is an interaction between radiations and living matter, there is always an *effect*. This effect can consist of purely physical or physicochemical phenomena without biological consequences: transitory phenomena followed by an almost immediate return to the initial state. However, it can also lead to more or less irreversible biological processes involving favorable, neutral, or noxious consequences to health.

Damage designates a noxious effect that causes a detectable impairment to the health of the exposed individual or to that of his or her progeny. Usually, this concept of damage has a qualificative meaning. When damage can be expressed quantitatively, for instance as the product of an incidence probability multiplied by the severity factor, this damage is sometimes called *detriment.*

The main objective of the present document is to define standards relating to the exposure of tissues or the whole body to NIR, in order to prevent damage to health. These standards can therefore be called *health protection standards* or exposure standards as opposed to standards for equipment, often referred to as *product performance standards* or *emission standards,* which should be established for the specifications and characteristics of NIR-emitting equipment to ensure safe operation. Equipment standards should be compatible with health protection standards.

For NIR, the health protection standards apply in general to the characteristic parameters of the radiation field at the point in space where an individual can be or is exposed. In this case, the quantities used are mainly:

- the power per unit area;
- the energy per unit area;
- the electric (or magnetic) field strength for electromagnetic radiation of certain frequencies;
- the sound pressure level for airborne ultrasound.

In some cases, they can also apply to the exchange of energy between radiation and matter. Then, the quantity used (with many restrictions) will be the energy or the power transferred to tissues per unit of mass.

1. The *power per unit area* (in free space) is defined by many terms. Unfortunately, many obstacles prevent the use of a single terminology. The following are used:

- power density;
- energy fluence rate;
- irradiance;
- acoustic intensity.

The International System (SI) unit used is the watt per square meter, $W \cdot m^{-2}$, and its most frequently used submultiples are the $mW \cdot cm^{-2}$ and the $\mu W \cdot cm^{-2}$.

2. The *energy per unit area* (in free space), that is, the product of the mean power density and the exposure time, is the *radiant exposure.* It is expressed in joules per square meter $J \cdot m^{-2}$, and is most often used for the exposure of skin and eyes to optical radiation.

3. Generally speaking, an electromagnetic wave is defined by its *electric* and *magnetic field strengths, E* and *H*, which are expressed in volts per meter ($V \cdot m^{-1}$) and amperes per meter ($A \cdot m^{-1}$), respectively.

When the distance from the source is great enough of the order of several wavelengths (far field conditions), the two fields are in phase and the *power density* is equal to the vector product of E and H.

Under these conditions, an inverse square law applies to the variation of power density with distance. The relation between E and H can be expressed by $E/H = 377\ \Omega$, with 377 Ω being the intrinsic impedance of air. Thus, the power density is proportional to E^2 or H^2 [$E^2/377$ or $H^2 \times 377$].

At low frequencies, the only practical way is to express exposure limits directly in terms of field strength ($V \cdot m^{-1}$, and $A \cdot m^{-1}$).

4. For airborne acoustic waves and particularly for ultrasound the limits apply to the *sound pressure level* (SPL). This is a pure number equal to:

$$10 \log \frac{I}{I_r} = \text{SPL}$$

where I is the acoustic intensity or power density for the wave studied and I_r the same quantity for the reference wave, or conventionally $I_r = 10^{-12}\ W \cdot m^{-2}$—that is, approximately the weakest intensity of audible sound that the human ear can perceive.

SPL is expressed in decibels (dB). As a result of the above relationship, doubling the intensity increases the SPL by three decibels (log 2 $\simeq 0.3$).

All the quantities listed above characterize the field at a point in free space where an individual might be subjected to whole-body or partial-body exposure. It is obvious that a quantity taking into account the presence of the human body and the interactions of radiation within the body and tissues would be much more expressive of the damage that might result.

For certain NIR (e.g., radiofrequencies), it was therefore found useful to introduce a quantity similar to the absorbed dose rate for ionizing radiation. This quantity, which has been called *specific absorption rate* or SAR, refers to the energy transferred to tissues per unit of time and of mass. It is expressed in watts per kilogram.

The SAR can be calculated and in some cases measured. It varies considerably with numerous parameters such as dielectric properties; even when the outside field is perfectly uniform, the SAR values can be different in the various tissues and in different parts of an organ. Resonant absorption conditions (macroscopic resonance) that depend greatly on the shapes and sizes of the bodies play a very important and complex role, even when only the mean values of SAR in a tissue or in the whole body are calculated.

All these difficulties in assessing SAR make it a parameter that, in spite of its considerable theoretical value, can only be used in practice in a limited number of cases.

LEVELS AND CONDITIONS OF HUMAN EXPOSURE

Except for exposure to optical radiation from the sun, it does not appear that man has been exposed in the past to high levels of natural NIR. Ultraviolet radiation, lasers, and ultrasound are increasingly used in research, industry, medicine, and even in consumer products. As regards radiofrequencies, for which a certain amount of numerical data is available, a low natural background is known to exist. Atmospheric electricity and emissions of electromagnetic waves by the sun and stars produce weak fields with a mean power density on the order of 10^{-8} mW · cm^{-2}.

Man-made sources that have increased in the past few decades (radio broadcasting; television; radars; and scientific, industrial, medical, and domestic uses of NIR-producing devices) not only lead to occupational exposure but also to exposure of the general public that is several orders of magnitude above the natural background. It is estimated that in some highly industrialized areas, the ambient power density at certain frequencies (predominantly at radio and television broadcast frequencies) is on the order of 10^{-3} mW · cm^{-2}, that is, a factor of 10^5 higher than the natural background. Although such an increase above the natural level constitutes a new element in the human environment, it is universally accepted that it does not produce an evident health risk.

Exposure sources are thus becoming considerably more numerous and consequently a need exists to promulgate a general protection policy for occupational and public exposure.

BIOLOGICAL EFFECTS

The varied biological effects that are due to the different types of NIR and the problems they create have been dealt with in greater detail in the IRPA/INIRC guidelines for each type of radiation. It is useful here to review their main characteristics.

Although the fundamental effect here is not that of ionization, the changes undergone by the molecules due to the various interactions may lead to functional and structural modifications, mutations, and cell death. Recovery will depend upon the extent and nature of the damage, and the final effect may either not be noticeable or may cause a detectable lesion.

The nature, the extent, and the physiological importance of the effects induced by the various types of NIR vary widely and depend upon a large number of factors, among which the following distinctions can be made:

- *parameters related to the incident radiation:* such as radiation frequency—or the wavelength or photon energy—that in particular will determine the penetration depth of the radiation in the tissues and can, in some cases, involve special phenomena such as resonance; the strength or power density of the field or beam; radiation emission characteristics (coherent or noncoherent, continuous, modulated, or pulsed); and in the latter case, the pulse durations and the interval between them;
- *parameters related to exposure conditions:* exposure duration and time distribution (continuous or intermittent exposure); orientation of the exposed individual within the radiation field; spatial distribution (whole-body or partial-body exposure);
- *parameters related to the biological characteristics:* especially the nature, the cellular, and even the molecular composition of the exposed tissues; their electric properties; certain physiological characteristics such as blood flow in the case of thermal effects, pigmentation, the size of the organ or of the individual; the functional importance of the damaged tissues or organs; complicated secondary effects.

In spite of the fact that for the past few years interest in the biological effects of NIR has increased and that many research projects have been undertaken in this field throughout the world, the precise role played by the different parameters, their interference, and especially their quantitative relationship with the biological effect produced are still not well known.

Concerning the noxious effects produced by NIR, a distinction can be made (as it can for ionizing radiation) between those for which the severity varies as a function of the exposure and which always appear when the exposure exceeds given values, the so-called *nonstochastic effects,* and those for which only the probability of occurrence—and not the severity—increases when the exposure increases, the so-called *stochastic effects.*

Certain nonstochastic effects are relatively well known, for example, erythema and skin burns, as well as ocular effects (conjunctivitis, retinal burns, cataracts) for optical radiation, and effects due to heating of the tissue (commonly called thermal effects) resulting from exposure to radiofrequencies (including microwaves) and to ultrasound. Thresholds for the occurrence of some of these effects have been determined. On the other hand, uncertainties still exist concerning the nonthermal effects that might be produced in particular by low-power density microwaves and by ultrasound. However, neither the induction mechanism of these nonthermal effects nor their possible pathological consequences on man have been clarified. It is indeed extremely difficult to extrapolate to man the results obtained on relatively small laboratory animals with respect to the physical aspect of interaction between the radiation field and the exposed individual as well as with respect to the biological effects produced.

Concerning the stochastic effects (cancer induction, genetic effects), even if it is known that they exist in some cases (skin cancer due to UV radiation), for the other types of NIR, research is still at a preliminary stage.

Finally, it should be noted that except for certain effects due to optical radiation, scant data are available on the quantitative relationships between exposure to the different types of NIR and body response.

GENERAL PRINCIPLES FOR PROTECTION AGAINST NIR

The analysis and synthesis of present knowledge on the pathological effects of NIR and the difficulties encountered in the monitoring of individual exposure have led the IRPA/INIRC to establish a protection doctrine based on the following principles:

1. Compliance with health protection standards as defined in this document ensures adequate protection for occupationally exposed workers and for members of the general public against the hazards that might result from NIR;
2. Compliance with health protection standards should be guaranteed as far as possible by the development and compliance with performance standards that apply to the design and construction of NIR-emitting devices;
3. When safety cannot be sufficiently guaranteed by construction (performance standards) due to the emitter's characteristics or to its use, appropriate operational protection measures must be applied in order to comply with health protection standards.

Health protection standards

These standards, such as the IRPA/INIRC guidelines, are intended to be applied to human exposure except for deliberate medical exposure (diagnostic or therapeutic).

They consist mainly of *limits* that the various quantities defined above should not exceed (except under certain circumstances specified in the text). The exact meaning of these quantities and the units involved are reiterated in detail in the following chapters for each type of radiation. The limit values should not be considered as a precise boundary between risk and no risk. In the light of present knowledge acquired through animal experimentation and human observation, they are thought to represent the best values to allow the general use of NIR under safe conditions. However, given the limited nature of available data and the progress that undoubtedly will be made concerning the pathological effects of NIR, it can be assumed that these limits might eventually be modified (either increased or lowered) if and when it becomes necessary.

Limits may be set for occupational exposure that differ from those for exposure of the population. Occupational exposure in principle concerns only adults (except for the embryo or fetus being carried by pregnant women). This exposure is limited by the working time and is subject to control. In addition, workers may benefit from medical surveillance and should be aware of the hazards involved and trained in appropriate protective measures. The population, on the other hand, includes children, senior citizens, and sick people, and exposure to certain types of NIR may take place 24 hours per day.

Device standards (product performance standards)

Because of the difficulties encountered in developing individual monitoring of exposure and therefore in ensuring compliance with health protection standards, it is all the more important to make provisions for the appropriate design and construction of the equipment and for the control and approval of the specifications.

The many industrial, scientific, medical, and home applications of high frequencies, microwaves, broad-beam optical radiation, lasers and ultrasound, lead to the design of widely varying devices. One of the first tasks is to classify them according to the power they develop and to the degree of safety for workers and for the public. For each category of apparatus, design and construction standards must be defined that take into account not only its intended use but also the necessary safety requirements. In particular, they should ensure that the device cannot materially deliver a needlessly high power (minimization of the emission), that whenever possible the direct radiation be confined or made inaccessible when the power density exceeds a given value and that the secondary radiation (scattered or leakage radiation) is reduced to a negligible value or even prevented in the proximity of the device.

The responsible authorities should see to it that standardized specifications, already elaborated in some cases by specialized international organizations, are made mandatory in each country by regulatory texts and that prototypes of equipment obtain the approval of specialized bodies before any mass-production and sale.

The conditions for use and necessary precautions to be taken for all equipment not having the intrinsic safety should be defined precisely and brought effectively by the manufacturer to the user's attention.

Operational protection

In most cases, technical protection measures applied to the source of radiation should be supplemented by operational protection measures adapted to each situation.

They cannot all be listed here due to their number and to the fact that they vary considerably depending upon the radiation, the type of emitter, and the circumstances. Some of the more frequently used measures are given below:

- selecting an appropriate site for the emitter (antennas of high-power radio and radar installations, long-distance lasers, etc.);
- delineating controlled, restricted, or forbidden areas;
- fitting the working place with special features (appropriate screens, automatic safety devices, etc.);
- identifying forbidden or controlled areas, operation of the devices, etc., such as by warning signs and lights;
- developing and implementing safety instructions for the use of apparatus at the various working places, including protection against the associated risks (electrical, chemical);
- using, under certain circumstances, individual protection equipment (special clothing, glasses, etc.);
- educating and training potentially exposed personnel;
- informing the public;
- monitoring exposure to NIR through the use of standardized, calibrated measurement equipment according to a procedure that is also standardized;
- conducting medical surveillance. When appropriate, medical surveillance adapted to the nature and importance of the risk to which workers are exposed can be necessary.

Health protection standards, device standards, as well as the operational protection measures should be the subject of a set of regulations and codes of practice. The IRPA/INIRC recommendations have been intended as guidelines for the national authorities who in their respective countries are responsible for the elaboration and implementation of such regulations.

REFERENCES

ILO85 International Labour Office, 1985, "Occupational hazards from non-ionising electromagnetic radiation," *Occupational Safety and Health Series No.* 53 (Geneva: ILO).

ILO86 International Labour Office, 1986, "Protection of workers against radiofrequency and microwave radiation: A technical review," *Occupational Safety and Health Series No.* 57 (Geneva: ILO).

UN79 United Nations Environment Programme/World Health Organization/International Radiation Protection Association, 1979, "Ultraviolet radiation," *Environmental Health Criteria No.* 14 (Geneva: WHO).

UN81 United Nations Environment Programme/World Health Organization/International Radiation Protection Association, 1981, "Radiofrequency and microwaves," *Environmental Health Criteria No.* 16 (Geneva: WHO).

UN82a United Nations Environment Programme/World Health Organization/International Radiation Protection Association, 1982, "Ultrasound," *Environmental Health Criteria No.* 22 (Geneva: WHO).

UN82b United Nations Environment Programme/World Health Organization/International Radiation Protection Association, 1982, "Lasers and Optical Radiation," *Environmental Health Criteria No.* 23 (Geneva: WHO).

UN84 United Nations Environment Programme/World Health Organization/International Radiation Protection Association, 1984, "Extremely low frequency (ELF) fields," *Environmental Health Criteria No.* 35 (Geneva: WHO).

UN87 United Nations Environment Programme/World Health Organization/International Radiation Protection Association, 1987, "Magnetic fields," *Environmental Health Criteria No.* 69 (Geneva: WHO).

CHAPTER 2

REVIEW OF CONCEPTS, QUANTITIES, UNITS, AND TERMINOLOGY FOR NON-IONIZING RADIATION PROTECTION

PREFACE

Scope of the report

NON-IONIZING radiation (NIR) is the term generally applied to all forms of electromagnetic radiation whose primary mode of interaction with matter is other than by producing ionization. Therefore NIR refers to electromagnetic radiation with wavelengths exceeding 100 nm equivalent to quantum energies below 12 eV, that is, encompassing the spectrum that includes all radiation sources whose frequencies are equal to or less than those of the near ultraviolet (UV). For the purpose of practical radiation protection, static electric and magnetic fields as well as energy transport through matter in the form of mechanical vibrations, such as ultrasound and infrasound, may also be considered as NIR.

Scientific, medical, industrial, and domestic uses of devices producing NIR are rapidly expanding in type and number, leading to a steady increase in the amount of NIR in man's environment and causing concern about potential health hazards to workers and to the general public from uncontrolled or excessive radiation exposure. Although these new technologies have been utilized significantly only in recent years and mainly in industrialized countries, it is likely that their wide applicability will result in their increasing dissemination all over the world with the potential that people in many countries will be exposed to NIR.

For purposes of health protection, electromagnetic NIR can be subdivided into a number of wavelength (λ) or frequency (ν) ranges:

- ultraviolet radiation, $100 \text{ nm} \leqslant \lambda \leqslant 400$ nm (optical radiation);
- visible radiation, $400 \text{ nm} \leqslant \lambda \leqslant 760$ nm (optical radiation);
- infrared radiation, $760 \text{ nm} \leqslant \lambda \leqslant 1$ mm (optical radiation);
- radiofrequency radiation including microwaves, $300 \text{ Hz} \leqslant \nu \leqslant 300$ GHz corresponding to $1000 \text{ km} \geqslant \lambda \geqslant 1$ mm;
- extremely low frequency (ELF) fields ($\nu \leqslant 300$ Hz), in practice mainly power frequencies of 50–60 Hz.

From a pragmatic point of view, magnetostatic and electrostatic fields are also dealt with in the framework of NIR.

Obviously, this is not the only classification available, and in practice various other classifications are used, according to particular needs. An international treaty, involving participants in the International Telecommunication Union, divides the range from 0–3 THz into 12 bands and allocates particular uses for certain bands (ITU81).

The field of NIR comprises also pressure waves such as ultrasound and infrasound including airborne ultrasound and infrasound that are on either side of the audible frequency range (20 Hz–20 kHz).

Protection against NIR is the subject of an increasing number of studies and publications, but the lack of an agreed upon terminology and the use of protection concepts that differ significantly between types of radiation and application makes it difficult to compare various studies and to compile data in a uniform way. Also, the development of regulations, standards, and public appreciation of the concepts of radiation protection are hampered by the lack of uniformity in terminology, quantities, and units.

In 1981 the International Non-Ionizing Radiation Committee (INIRC) of the International Radiation Protection Association (IRPA) set up a working group with the task to prepare the present report for workers in the field of NIR protection. The group consisted of B. Bosnjakovic (Chairman; Radiation Protection Directorate, Ministry of Housing, Planning and Environment, Rijswijk, The Netherlands), D. Harder (Institut für Medizinische Physik und Biophysik der Universität, Göttingen, Federal Republic of Germany) and D. Sliney (Laser Microwave Division, U.S. Army Environmental Hygiene Agency, Aberdeen Proving Ground, Maryland, U.S.A.). The final report was adopted by the INIRC in 1984.

During the preparation of this document the composition of the IRPA/INIRC was as follows:

H. P. Jammet, Chairman (France)
B. F. M. Bosnjakovic (Netherlands)
P. Czerski (Poland)
M. Faber (Denmark)
D. Harder (Federal Republic of Germany)
J. Marshall (United Kingdom)
M. H. Repacholi (Australia)
D. H. Sliney (U.S.A.)
J. C. Villforth (U.S.A.)
A. S. Duchêne, Scientific Secretary (France)

The main aim of the report is to provide an inventory of concepts, quantities, units, and terminology currently used for purposes of NIR protection. A systematic classification and comparison of these quantities is also given, and in particular the concepts used to quantify exposure limitations and radiation protection standards are summarized. To a limited degree, the report addresses the question whether and how an improved uniformity and harmonization of quantities and units can be achieved in the field of NIR. Throughout the report, the approach is pragmatic; account being taken as much as possible of existing practices in science and standardization.

The present report is regarded as a first step towards a more permanent recommendation on quantities and units for NIR protection. Comments are invited from individual scientists as well as national and international organizations.

Current standardization activities in the field of radiation

In addition to legislative bodies and regulatory authorities, there are other bodies actively concerned with radiation standardization. Three types of organizations and institutions are involved.

1. Governmental and intergovernmental, such as CGPM (Conférence Générale des Poids et Mesures) and CIPM (Comité International des Poids et Mesures), which is the executive committee serving CGPM. The CIPM supervises the operations of the BIPM (Bureau International des Poids et Mesures), which was founded in 1875. Most states have corresponding national governmental institutions; for example, NBS (National Bureau of Standards) in the United States. OIML (Organisation Internationale de Métrologie Légale) is another intergovernmental organization. There is an official agreement (since 1975) between OIML and the United Nations (UN) that guarantees the necessary coordination with the interested UN agencies, such as the World Health Organization (WHO) and the UN Environment Programme (UNEP).

2. Nongovernmental organizations with a broad scope of standardization activities: ISO (International Organization for Standardization), ITU (International Telecommunication Union), and IEC (International Electrotechnical Commission). There are corresponding national standardization institutions and organizations.

3. Specialized international associations and committees, representing professional and scientific groups active in various fields of science and technology. Those relevant for quantities and units in the NIR field include:

- AIP (Association Internationale de Photobiologie), formerly CIP (CIP54);
- CIE (Commission Internationale de l'Eclairage) (CIE70);
- ICRU (International Commission on Radiation Units and Measurements) (ICRU80);
- IRPA;
- IUPAP, IUPAC (International Union of Pure and Applied Physics, International Union of Pure and Applied Chemistry);
- URSI (Union Radioscientifique Internationale) Commissions A and B.

All of these organizations have contributed in one way or another to standardization in the field of radiation, some of them for many decades. Mutual contacts exist, sometimes through official channels, but mainly through personal contacts.

One intention of the present document is to make use of the results, the experience, and the knowledge collected by the above-mentioned bodies, thereby serving radiation protection standardization in the field of NIR.

QUANTITIES AND UNITS

The aim of this section is to present an inventory of quantities and units currently used in the field of NIR.

Physical quantities are used to describe and characterize physical phenomena in a quantitative way. For the purpose of radiation protection, physical quantities are needed (a) to describe sources and fields of radiation as well as (b) the interaction of radiation with matter. Some quantities have a special significance because they may be needed (c) to describe the exposure of the human body to NIR ("dosimetric quantities"); an important application of dosimetric quantities is in setting exposure limits.

A certain value of a physical quantity (e.g., electric current) is usually expressed as a multiple of a chosen *unit* (e.g., ampere). The use of the International System of Units (SI) (BIPM73) is in principle generally accepted today. SI units are divided into three classes—base units, derived units, and supplementary units. The base units are metre, kilogram, second, ampere, kelvin, mole, and candela for the quantities length, mass, time, electric current, thermodynamic temperature, amount of substance, and luminous intensity, respectively. The current supplementary units are radian and steradian for the quantities plane angle and solid angle, respectively.

Derived units are formed by combining base units and/or supplementary units according to the algebraic relations linking the corresponding quantities. In this report, the use of SI units is recommended for the whole field of NIR. Exceptions are explicitly mentioned, if necessary.

A broad and general approach to standardization has been chosen by the International Organization for Standardization (ISO). Typical radiation quantities appear in a number of ISO International Standards included in ISO82, such as those concerning the quantities and units of:

- periodic and related phenomena (ISO 31/2-1978);
- electricity and magnetism (ISO 31/5-1979);
- light and related electromagnetic radiation (ISO 31/6-1980);
- acoustics (ISO 31/7-1978);
- nuclear reactions and ionizing radiation (ISO 31/10-1980).

Many quantities and units recommended in the *ISO Standards Handbook No. 2* have been adopted. Since ISO recommendations are not written exclusively for radiation protection purposes, a selection has been made that excludes certain quantities of little conceivable interest for radiation protection. Some other quantities, important for radiation protection but not contained in ISO recommendations, have been added. In some cases definitions and remarks have been modified. In particular, completeness has been sacrificed for the sake of clarity of essential details, and in addition mathematical formulation was kept to a minimum.*

The compilation should not be considered exhaustive nor definitive, and it can not re-

* Material from Tables 2, 5, 6 and 7 of ISO 31/2, 31/5, 31/6 and 31/7 is reproduced with the permission of the International Organization for Standardization (ISO). Copies of the complete standards can be obtained from the ISO Central Secretariat, 1, rue de Varembé, 1211 Geneva 20, Switzerland, and from any ISO member body.

place existing, more specialised recommendations (see e.g., [NCRP81], pertaining to radiofrequency and microwave radiations from 300 kHz to 300 GHz).

The inventory of useful quantities and units in the field of NIR is tabulated as Appendix 1. The material is subdivided into the following sections:

1. Periodic and radiation phenomena;
2. Electromagnetic radiation and fields;
3. Optical radiation;
4. Ultrasound.

The presentation of the material follows closely the format chosen in the *ISO Standards Handbook No. 2,* "Units of Measurement" (ISO82), that is, in two parallel lists, one pertaining to quantities, the other to the corresponding units. The lists are self-explanatory and contain some brief conceptual definitions. It is assumed that the reader is familiar with the basic physical concepts and with the general quantities and units used in physics. No attempt was made to give a coherent, textbook-like introduction into the underlying theoretical or experimental fundamentals.

If more than one symbol is listed for the same quantity, the first symbol is generally preferred. It is unavoidable that there are situations where the same mathematical symbol is used to designate different physical quantities. SI units are used throughout, including prefixes indicating decimal multiples or submultiples of units (Appendix 2). Prefixes other than those listed are not excluded, for example, W/cm^2 may be used as well as W/m^2. For some of the older non-SI units, conversion factors are stated.

However, different names for the same or similar quantity are used, either within the same International Standard (e.g., $\omega = 2\pi f$ is called angular frequency, circular frequency, or pulsatance in ISO 31/2) or in different International Standards (e.g., ISO 31/10 uses the expressions energy flux density and energy fluence rate, in ISO 31/6 the term radiant energy fluence rate is used, whereas in ISO 31/5 the magnitude of the Poynting vector is used to characterize energy flux density). More examples can be found, and this diversity is reflected in Appendix 1.

CLASSIFICATION AND COMPARISON OF QUANTITIES

Taking into consideration the practical needs of radiation protection, the quantities compiled in Appendix 1 may be classified according to three criteria:

1. The physical characteristics of the radiation field, taking sources and receivers of radiation into account (radiometric quantities);
2. The interactions of NIR with matter (e.g., interaction coefficients);
3. Quantities adequate for the specification of exposure of biological objects to NIR (dosimetric quantities). (This classification will be discussed in more detail below.)

It is important to note that the biological object has to be accounted for in all three categories, because:

- its presence as a receiver modifies the radiation field and can be the cause of a difference between the "perturbed" and the "unperturbed" field;
- the biological object itself undergoes physical interactions that form the basis for the subsequent biological effects; and
- the biological responses, together with other practical constraints, influence the choice of quantities adequate to specify exposure and limits of exposure.

The concepts and quantities for ionizing radiation may be divided into the following categories (ICRU80):

- "radiometry," which deals with quantities associated with the radiation field;
- "interaction coefficients," which deal with quantities associated with the interaction of radiation and matter; and
- "dosimetry," which deals with quantities that are generally products of quantities in the first and second categories.

(The category "Radioactivity" also considered by ICRU is not required in the present context.)

The classification of the ICRU is close to that chosen above for NIR. Thus, Youmans and Ho (Yo75, p. 313) give the following conceptual base for dosimetric quantities in the context of radiofrequency electromagnetic radiation:

> "A transport equation describes the flow of radiation in terms of the strength of the source, the absorption and scattering properties of the matter present, and the quantities which characterize the radiation field. In this sense, absorbed dose may be considered to be the image of exposure under a transformation (physical operator) which accounts for the absorption, scattering and geometric properties of an irradiation condition, such that absorbed energy (an absorbed dose parameter) is the work done on matter by a field of radiation (an exposure parameter)."

However, the formal analogy among the three categories listed above and the three categories given by the ICRU should not be taken as proving that dosimetric quantities in the NIR field *must* necessarily be formed as products of radiometric quantities and interaction coefficients. There is a great diversity of physical interaction phenomena and of biological response mechanisms, as well as large differences in measuring techniques and, in some cases, a considerable lack of knowledge, which have in practice led to the selection of special dosimetric quantities for NIR.

At this point, a more general remark is necessary. Harmonization of quantities and units is desirable both from a systematic and a practical point of view (Ru77). However, there are limitations and obstacles that limit unification and that have to be taken into account in practical situations:

- the larger the field to be unified, the higher is the necessary degree of abstraction;
- there are divergent traditions and professional interests; and
- practical considerations of different measurement techniques and different biological mechanisms must be observed.

Quantities for the characterization of sources and fields

Radiometric quantities. In Tables 1 and 2 some essential examples of quantities have been extracted from Appendix 1 in order to show the correspondence between the various non-ionizing and ionizing radiations with regard to their grouping into radiometric quantities. These tables also serve the purpose of clarifying the differences existing in terminology between corresponding quantities for various kinds of radiation, to compare radiometric and photometric quantities and to give a comparative survey of the "area element dA" which has to be considered in various definitions.

Table 1 contains the radiometric quantities and units used in the various subfields of NIRs. Eight generic terms (energy; energy per time; energy per area; energy per volume; energy per time and solid angle; energy per time, area, and solid angle; and energy per area and solid angle) are considered for electromagnetic radiation (with emphasis on radiofrequency), for optical radiation (including ultraviolet and infrared radiation), and for sound (with emphasis on ultrasound).

Ionizing radiation was also included, to demonstrate the similarities of approach and the degree of standardization in this field.

The columns are arranged to show the increasing complexity of the relationship between the energy and the variables time, surface, volume, and solid angle.

The quantity energy and its time derivative, energy flux or power, (columns 2 and 3 of Table 1), are most frequently applied to characterize a source. In some fields they may also be termed "radiant energy" and "radiant power." For ultrasound, the prefix "acoustic" is added. The basic concept and definition of these quantities is uniform throughout the fields of non-ionizing and ionizing radiations.

Spatial energy density (energy per volume) and energy transport through space (energy per area or energy per time and area) at any given point of space in a medium are described in columns 4, 5, and 6 of Table 1.

Since all radiometric quantities are applicable to multidirectional radiation, great care must be taken in specifying the orientation of the surface of a source or receiver in relation to the radiation field (Mo79). Within the existing framework of quantities, three approaches are being used and must be carefully distinguished in practical applications:

TABLE 1. *Synopsis of comparable radiometric quantities (numbers in parentheses refer to item number of Appendix 1)*

Type of radiation	Generic term: Energy (J)	Energy/time (W)	Energy/area (J/m²)	Energy/volume (J/m³)	Energy/(time · area) (W/m²)	Energy/(time · solid angle) (W/sr)	Energy/(time · area · solid angle) (W/m²sr)	Energy/(area · solid angle) (J/m²sr)
Ionizing radiation	Radiant energy R	Energy flux $\dot{R} = \frac{dR}{dt}$	Energy fluence * $\Psi = \frac{dR}{dA}$		Energy fluence rate, energy flux density * $\psi = \frac{d\Psi}{dt} = \frac{d^2R}{dA\,dt}$		Energy radiance ** $r = \frac{d\psi}{d\Omega} = \frac{d^2\Psi}{dt\,d\Omega} = \frac{d^3R}{dA\,dt\,d\Omega}$	
Radiofrequency electromagnetic radiation	Radiant energy Q (2.32)	Radiant power, radiant energy flux $P = \frac{dQ}{dt}$ (2.33)		Electromagnetic energy density $w = \frac{dQ}{dV}$ (2.34)	Surface power density, Energy flux density * $\Psi = \frac{d^2Q}{dA\,dt}$ (2.36)	Power per solid angle $I = \frac{d^2Q}{dt\,d\Omega}$ (2.37)		
Optical radiation (radiometric quantities)	Radiant energy Q (3.1)	Radiant power, radiant energy flux $P = \frac{dQ}{dt}$ (3.3)	Radiant exposure *** $H = \frac{dQ}{dA}$ (receptor surface) (3.18)	Radiant energy density $w = \frac{dQ}{dV}$ (3.7)	Radiant energy fluence rate* $\phi = \frac{d^2Q}{dA\,dt}$ (3.5) Radiant exitance *** $M = \frac{d^2Q}{dA\,dt}$ (source surface) (3.14) Irradiance *** $E = \frac{d^2Q}{dA\,dt}$ (receptor surface) (3.16)	Radiant intensity $I = \frac{d^2Q}{dt\,d\Omega}$ (3.9)	Radiance ** $L = \frac{d^2Q}{dA\,dt\,d\Omega}$ (3.11)	Time-integrated radiance ** $\Lambda = \frac{d^2Q}{dA\,d\Omega}$
Ultrasound	Acoustic energy Q	Acoustic power $P = \frac{dQ}{dt}$ (4.9)		Acoustic energy density $w = \frac{dQ}{dV}$ (4.8)	Acoustic intensity * $I = \frac{d^2Q}{dA\,dt}$ (4.10)			

* The area dA is taken as the cross section of a sphere on which the radiation is incident at the point under consideration.
** The area dA is taken as perpendicular to the direction of the radiation.
*** The normal on area dA forms an angle θ with the direction of the radiation.

TABLE 2. *Comparison of radiometric and photometric terms (numbers in parentheses refer to item numbers of Appendix 1)*

Radiometric term		Photometric term	
Quantity	Unit	Quantity	Unit
Radiant energy (3.1)	Joule (J)	Quantity of light (3.24)	lumen-second (lm · s) (talbot)
Radiant energy density (3.7)	Joule per cubic metre ($J \cdot m^{-3}$)	Luminous density *	lumen-second per cubic metre ($lm \cdot s \cdot m^{-3}$) (talbot per cubic metre)
Radiant power (radiant energy flux) (3.3)	Watt (W)	Luminous flux (3.22)	lumen (lm)
Radiant intensity (3.9)	Watt per steradian ($W \cdot sr^{-1}$)	Luminous intensity (3.21)	lumen per steradian (cd or $lm \cdot sr^{-1}$)
Radiant exitance (3.14)	Watt per square metre ($W \cdot m^{-2}$)	Luminous exitance (3.26)	lumen per square metre ($lm \cdot m^{-2}$)
Radiant energy fluence rate (3.5)	Watt per square metre ($W \cdot m^{-2}$)	Luminous flux density *	lumen per square metre ($lm \cdot m^{-2}$)
Radiance (3.11)	Watt per steradian and per square metre ($W \cdot sr^{-1} \cdot m^{-2}$)	Luminance (3.25)	candela per square metre ($cd \cdot m^{-2}$)
Irradiance (3.16)	Watt per square metre ($W \cdot m^{-2}$)	Illuminance (3.27)	lumen per square metre ($lm \cdot m^{-2}$), lux (lx)
Radiant exposure (3.18)	Joule per square metre ($J \cdot m^{-2}$)	Light exposure (3.28)	lux-second (lx · s)

* Stated here for comparison only. Not in Appendix 1 Table 3.

1. *Area element perpendicular to the direction of radiation.* For quantities that refer either to a unidirectional beam or to that portion of a multidirectional radiation field that is confined within a small solid angle of given direction, the area element, dA, is taken as perpendicular to this direction. This approach is used in practice for the quantities energy radiance (ionizing radiation), radiance, and time-integrated radiance (optical radiation).

2. *Cross-sectional area of a sphere.* In the definitions of energy fluence and energy fluence rate (ionizing radiation), surface power density or energy flux density (electromagnetic radiation), radiant energy fluence rate (optical radiation), and acoustic intensity (ultrasound), the reference surface is conceived as the cross-sectional area, dA, of a small sphere surrounding the point under consideration. A sphere has the property of presenting the same cross-sectional area perpendicular to each direction of incidence. As these quantities form integrals or sums over the contributions from all directions of incidence, the independence of the sphere cross section from the direction of incidence means that equal weight is attributed to each direction.

3. *Area element not perpendicular to the radiation.* For exposure to optical radiations, the penetration depth in matter is generally very small. Therefore, the emission from a radiating body is characterized by the amount of energy radiated from a given surface area, and the degree of biological effect is often determined by the amount of radiant energy incident on a given surface area. The relevant quantities are therefore related to the area element, dA, of an emitting or receiving surface, whose normal may form any angle, θ, with the direction of radiation.

The generic term *energy per area* is used in two ways:

1. The amount of energy passing per unit cross section of a small sphere ("energy fluence" for ionizing radiation, "fluence" for optical radiation, e.g., in photobiology); and
2. The amount of energy passing per unit area through a receiving surface ("radiant exposure" for optical radiation, traditionally called "dose" in photobiology).

For the difference between fluence and dose in photobiology, see, e.g., Ph84.

The term *energy per time and area* is used in three ways:

1. Radiant power per unit cross section of a small sphere (energy fluence rate, radiant energy fluence rate, energy flux density, surface power density, acoustic intensity);
2. Radiant power per unit area of the source surface ("radiant exitance" for optical radiation); and
3. Radiant power per unit area of a receiving surface (optical "irradiance").

Sources of radiation with an angular dependence can be characterized by the radiant power per unit solid angle ("radiant intensity"), the radiant power per unit area and unit solid angle ("radiance"), and the radiant energy per unit area and unit solid angle ("time-integrated radiance"), as shown in columns 7, 8, and 9 of Table 1. These quantities are especially useful if the source may be imaged by the eye. It should be noted that the quantity radiance is invariant throughout an optical beam; this is equally applicable for the most general descriptions of the unattenuated radiation either emitted from a source, passing through an arbitrary surface in a medium, or incident upon a receiving surface. (If the optical beam passes through vacuum, the radiance L is constant. If it passes through a medium with refractive index n, the quantity L/n^2 is constant).

It can be concluded that only three quantities are reserved for exclusive characterization of either a radiation source or an irradiated surface: *radiant exitance* (for sources), and *irradiance* and *radiant exposure* (for receiving surfaces). This exclusive use is indicated by the linguistic appearance of these terms.

Radiometric and photometric quantities. A comparison between "radiometric" and "photometric" quantities for optical radiation is provided in Table 2 where the former, already discussed in the context of Table 1, are listed in parallel to their photometric analogs. As noted in the remarks on quantities 3.29–3.31 in Appendix 1, the relationship between the radiometric quantities (index e) and the photometric quantities (index v) is given by the relation

$$\Phi_{v\lambda} = K_m V(\lambda)\Phi_{e\lambda}, \qquad [1]$$

where

$\Phi_{v\lambda}$ = spectral luminous flux,

$\Phi_{e\lambda}$ = spectral radiant power,

K_m = maximum spectral luminous efficacy,

and

$V(\lambda)$ = spectral luminous efficiency.

Photometric quantities are not only based on radiation properties, since they represent radiometric quantities weighted with the response function $V(\lambda)$ (see equation [1]) relating to the light-adapted eye. They have been shown to be useful concepts in the evaluation and control of hazards associated with specific types and uses of lamps and visual displays (S180).

Radiant power with its analog luminous flux, radiant intensity with luminous intensity, and radiant exitance with luminous exitance, are the pairs of quantities typical for the characterization of optical sources.

Radiance and luminance, radiant energy fluence rate and luminous flux density, and radiant energy density and luminous density are of general use for any position in the optical radiation field. The pairs of quantities determined to quantify the optical irradiation of surfaces are irradiance and illuminance, and radiant exposure and light exposure.

Limitations. Energy, power, and the derived radiometric quantities are concepts that become less satisfactory as the wavelength of the radiation increases. One practical reason for this can be given as follows:

In an idealized exposure condition, sufficiently remote from the source, the ratio of the electric to the magnetic field strength is $E/H = 377\Omega$. For coherent sources such far-field conditions usually apply for distances greater than $2a^2/\lambda$ where a is the dimension of the coherent source, say an antenna, and λ is the wavelength. The energy flux density EH therefore can be determined as $E^2/(377\Omega)$ or as $H^2 \cdot 377\Omega$. At distances less than $2a^2/\lambda$, the inverse square law does not apply, and the ratio E/H can differ substantially from 377Ω.

Instruments calibrated in units of power, which in reality respond to E, will become increasingly inaccurate at closer distances. This problem begins at microwave frequencies, and becomes even more important for lower radiofrequencies and extremely low frequencies. Under such conditions, a sufficiently general specification of the electric and magnetic field strengths, *E*, *H*, of the electromagnetic wave at each location of interest is required.

Since incident radiation may be reflected, transmitted, scattered, and absorbed by biological structures, a distinction must be made between (a) the radiation field existing in space in the absence of the exposed body ("free field") and (b) the radiation field in the presence of an exposed body. In the second case it is necessary to distinguish further between the radiation field within and in the vicinity of the exposed body.

When the wavelength is comparable with the dimensions of the body, it may not be possible to define a penetration depth, and it might be preferable to speak of the field strengths inside and outside.

Quantities for the characterization of interaction processes

General. The concepts of scattering, attenuation, transmission, reflection, refraction, and diffraction are well known from the physical theories describing radiation phenomena in media and at their mutual boundaries. The underlying electromagnetic and mechanical properties of matter that can be defined without explicit reference to incident radiation (such as conductivity, permittivity, permeability, polarization) are to some extent included in Appendix 1 (Table 2).

Although biological interactions of NIR can be thought of in terms of quantum-mechanical effects, the use of specific quantum-mechanical quantities and units for radiation protection purposes is not considered.

The quantities describing interaction processes are summarized in Table 3. Whereas attenuation, absorption, and scattering can be defined for any point in a medium, reflection is a quantity that relates only to a boundary between two media. For reasons of simplicity, refraction is put into the same column as reflection.

Diffraction, being determined essentially only by the geometry of bodies and tissues, is not treated in this context. However, diffraction phenomena can be of great importance for radiation protection, for example, if the dimensions of bodies or layers—including the dimensions of the radiation source—are comparable with the wavelength of incident radiation. The quantities describing the specific influence of diffraction and interference effects in the context of radiation protection are the radiometric quantities (Table 1).

In Table 3, the quantities describing attenuation, absorption, scattering, and reflection/refraction are compared for ionizing radiation, electromagnetic radiofrequency radiation, optical radiation, and ultrasound, similar to the comparison made in Table 1. In the description of interaction processes, analogies between ionizing and non-ionizing radiation are limited, due to the different nature of the underlying physical interaction processes. The extremely short wavelengths involved in the interactions of ionizing radiation with matter cause interactions at the atomic and nuclear level. In the case of NIR, interest and importance generally shift to "bulk" properties such as the dielectric constant, conductivity, compressibility, and mean density, as well as reflection and interference phenomena.

Consequently, material and geometric properties of the target biological structures, especially the interfaces, become dominant factors.

Attenuation concepts. The mass attenuation coefficient, μ/ρ, for uncharged ionizing particles (ICRU80), is the quotient of $\mathrm{d}N/N$ by $\rho \mathrm{d}l$, where $\mathrm{d}N/N$ is the fraction of particles that experience interaction in traversing a distance $\mathrm{d}l$ in a material of density ρ. For radiofrequency waves, the term attenuation is applicable both to the amplitude (e.g., of the electric field strength) and to the power. For optical radiation, attenuation refers only to power whereas for ultrasound it is also applied to the amplitude of the acoustic pressure.

Particle description. Since the particle description of radiation has limited value for the characterization of NIR interaction processes, the quantities used for ionizing corpuscular radiation such as mass stopping power, linear stopping power, linear energy transfer, and many others are not applicable to NIR. Whereas photon energy and photon flux can be useful in photobiological studies, these

TABLE 3. *Selected quantities describing interactions of radiations with matter (numbers in parentheses refer to item numbers of Appendix 1)*

Radiation	Interaction process: Attenuation	Absorption	Scattering	Reflection and related phenomena
Ionizing radiation	Linear attenuation coefficient μ (m^{-1}) Mass attenuation coefficient μ/ρ (m^2/kg) Half-value thickness $d_{1/2}$ (m)	Linear energy absorption coefficient μ_{en} (m^{-1}) Mass energy absorption coefficient μ_{en}/ρ (m^2/kg)	Atomic cross sections for Compton scattering, Rayleigh scattering, elastic nuclear scattering, electron-electron scattering (m^2)	Backscattering factor
Radiofrequency electromagnetic radiation	Transmission factor τ (2.39) Attenuation coefficient α (2.40) (m^{-1}) Depth of penetration $1/\alpha$ (m) (2.40)	Dissipation factor (loss tangent) tan $\sigma = \varepsilon_r/\varepsilon_i$ Absorption coefficient α_a (m^{-1}) (2.40)	Scattering cross section (m^2) Scattering coefficient α_s (m^{-1}) (2.40)	Reflection factor Γ (2.39)
Optical radiation	Spectral transmittance $\tau(\lambda)$ (3.34) Spectral linear attenuation coefficient $\mu(\lambda)$ (m^{-1}) (3.36)	Spectral absorptance $\alpha(\lambda)$ (3.32) Spectral linear absorption coefficient $a(\lambda)$ (m^{-1}) (3.37) Spectral molar absorption coefficient $\kappa(\lambda)$ (m^2/mole) (3.38)	Spectral radiance factor $\beta(\lambda)$ (3.35)	Spectral reflectance $\rho(\lambda)$ (3.33) Spectral refractive index $n(\lambda)$ (3.39)
Ultrasound	Pressure transmission factor t (4.21) Intensity transmission factor T (4.23) Amplitude attenuation coefficient α (m^{-1}) (4.24) Level attenuation coefficient α' (dB/m) (4.26)	Amplitude absorption coefficient α_a (m^{-1}) (4.25) Level absorption coefficient α_a' (dB/m) (4.27)	Amplitude scattering coefficient α_s (m^{-1}) (4.25) Level scattering coefficient α_s' (dB/m) (4.27)	Pressure reflection factor r (4.20) Intensity reflection factor R (4.22) Intensity backscattering coefficient B (m^{-1} sr^{-1}) (4.28)

quantities are not used for radiation protection purposes.

Coefficients. The terminology used to describe quantitatively the different interaction parameters shows a rather high degree of uniformity. The term "coefficient" (as for scattering coefficient, absorption coefficient, etc.) is generally used for the relative decrease of a radiometric quantity due to an interaction process (scattering, absorption) during passage through an infinitesimal layer of a medium, divided by the thickness of that layer. The unit of an interaction coefficient is the reciprocal metre (m^{-1}). It should be noted that the interaction coefficient can be defined either in terms of amplitude decrease (amplitude coefficient) or power decrease (power coefficient). If amplitude level and power level, respectively, are considered instead of amplitude and power, the corresponding level coefficient is measured in the unit dB/m.

In contrast to interaction coefficients, the term "factor" is reserved for dimensionless ratios of radiometric quantities, either amplitude or power, to describe phenomena at boundaries or finite layers: reflection factor, transmission factor, and backscattering factor. These quantities have obvious practical importance for radiation protection. It should be noted that for dimensionless ratios of radiometric quantities other names are sometimes used, such as spectral absorptance, reflectance, and transmittance, which are used in optics. It should be recognized that these terms apply inherently to idealised conditions, assuming a plane wave and an infinitely extended medium. Important modifications may be necessary for special situations like collimated beams, media of finite dimensions, etc.

Quantities for the specification of exposure to NIR (dosimetric quantities)

In a broad sense, the term "dosimetry" is used to quantify an exposure to radiation. Quantitative descriptions of an exposure to radiation, for the purpose of formulating protection standards and exposure limits, require the use of adequate quantities. "Adequate" means that the quantities should represent as well as possible, those physical processes that are closely linked to the biological effects of radiation.

Whether such quantities can simply be related to the amount of energy imparted to tissue such as in the case of ionizing radiation, or whether other quantities such as field strengths are more appropriate, is one of the most important questions in NIR dosimetry.

Since many specialists in radiation protection have a strong background in the dosimetry of ionizing radiation, it is useful to draw comparisons between dosimetric concepts used for NIR and ionizing radiation. Dosimetric concepts for *ionizing radiation* have been developed since the beginning of this century and have resulted in a high degree of sophistication. The results of research done in dosimetry and microdosimetry of ionizing radiation have found their way into the field of standardization, for example, ICRU concepts like energy imparted, lineal energy, and absorbed dose. For the specification of exposure to radiation, the most important quantity is absorbed dose (ICRU80), measured in grays, with all its variants. Absorbed dose can be modified by suitable factors, such as Q (quality factor) to obtain quantities more representative of biological effects such as dose equivalent and effective dose, measured in sieverts. Some definitions, such as "kerma," are unique to ionizing radiation dosimetry since they describe the release of charged ionizing particles in matter.

In the case of NIR, different characteristics of physical interaction mechanisms, measurement conditions and techniques, as well as differences in (and the limited knowledge of) biological response mechanisms have led to a diversity of quantities used for the specification of exposure.

In general, it can be stated that, across the NIR spectrum, the temporal characteristics of exposure are of critical importance, and the contributions of ambient factors such as temperature must be taken into account. For example, if the emphasis is on the limitation of thermal effects, many data in the radiofrequency region at present seem to support the introduction of exposure limits that are based on the *rate* of energy deposition, as opposed to absorbed dose in the case of ionizing radiation (see also NCRP81). Also thermal stress due to infrared radiation obviously requires a special treatment. Acoustic fields represent another example where quantities other than cumulative energy deposition may be useful to specify exposure limits.

In the following discussion of the quantities that have been selected for the specification of exposure and of limits of exposure to NIR, these viewpoints will be applied with respect to the various radiations. A compilation of the quantities used for expressing limits of exposure is given in Table 4, and may be compared with the selection made for ionizing radiation.

In the *radiofrequency spectrum,* the use of energy absorption as a parameter to characterize exposure is highlighted by the quantity specific absorption rate (SAR), (unit: W/kg). Like absorbed dose rate in the case of ionizing radiation, SAR is a quantity that is defined locally (for a small amount of tissue), and the concept is extended to include spatial peak SAR as well as temporal peak SAR, and SAR averaged over part of an organ or the whole body. The concept of SAR corresponds mainly, but not exclusively, to the thermal mechanism of biological action (Chapter 5; UN81).

For practical purposes of radiation protection, exposure limits for radiofrequencies above 10 MHz are given in terms of the quantity "surface power density" in free space, expressed in W/m^2, with higher limits allowed for exposures of limited duration. The limitations of this practice have been criticized, among others, by Youmans and Ho (Yo75). For this reason, the IRPA guidelines for exposure limits for radiofrequency radiation give "basic limits" in terms of SAR and "derived limits" in terms of surface power density (Chapter 5).

For radiofrequencies below 10 MHz, basic limits are expressed in terms of the "effective electric field strength," E_{eff}, and the "effective magnetic field strength," H_{eff}. Since, in the near field, the phase relationship between the directional components of a field strength is usually unknown, the "effective field strength" is obtained by adding the squares of the amplitudes of the vertical and the horizontal components and taking the square root of this sum. The effective field strengths are deter-

TABLE 4. *Selected quantities for the specification of limits of exposure to non-ionizing radiation (dosimetric quantities)*

Radiation type	Field parameters	GENERIC TERMS					
		Energy / area	Energy / time · area	Energy / time · area · solid angle	Energy / area · solid angle	Energy / mass	Energy / mass·time
Ionizing radiation						Absorbed dose (Gy) Dose equivalent * (Sv)	Absorbed dose rate (Gy/s) Dose equivalent rate * (Sv/s)
Radiofrequencies	Effective electric field strength (V/m) (2.5) Effective magnetic field strength (A/m) (2.15)		Surface power density (W/m^2) (2.36)			Specific absorption (J/kg) (2.41)	Specific absorption rate (W/kg) (2.42)
ELF	Electric field strength (V/m) (2.5) Magnetic field strength (A/m) (2.15)						
Optical		Radiant exposure (J/m^2) (3.18)	Irradiance (W/m^2) (3.16) Effective irradiance (W/m^2)	Radiance ($W \cdot m^{-2} \cdot sr^{-1}$) (3.11)	Time-integrated radiance ($J \cdot m^{-2} \cdot sr^{-1}$) (3.12)		
Ultrasound			Acoustic intensity (W/m^2) (4.10)				
Ultrasound (airborne)			Acoustic pressure level (dB) (4.16)				

* Quantity involving a biological weighting factor.

mined in free space, that is, in the absence of an absorbing or scattering body.

For the reasons given in the section on quantities for the characterization of sources and fields, p. 12, the quantities surface power density, specific absorption rate, and specific absorption are not practical for very long wavelengths, such as in the case of *ELF fields.* Due to the poor understanding of interaction mechanisms, the dosimetric concepts are not fully developed for ELF fields. At the moment, exposure conditions are often quantified by a statement of the unperturbed external electric and magnetic field strengths and the duration of exposure.

In the case of *optical radiation,* the situation is more complicated. Generally, exposure of the eye and skin is specified in different formats. Skin exposure is characterized by the incident energy per area (not necessarily orthogonal to the direction of incidence), expressed in J/m^2 and termed "radiant exposure" whereby an interaction parameter, such as penetration depth, may or may not be specified. In photobiology the traditional term "dose" has been in widespread use for this quantity. In the context of skin irradiation, exposure is sometimes quantified in multiples of the minimal erythema dose (MED) for an individual. The MED, expressed also in J/m^2, depends on the wavelength. These multiples correspond formally to the quantification of exposure to ionizing radiation by means of the quantity "dose equivalent."

For the characterization of ocular exposure, several radiometric quantities may be used: radiant exposure (J/m^2), irradiance (W/m^2), radiance ($W/m^2 \cdot sr$), and time-integrated radiance ($J/m^2 \cdot sr$), the two latter quantities being used for extended radiation sources. Exposure duration has to be specified, and the statement of spectral transmittance can also be useful. Since all photometric quantities are equivalent to radiometric quantities weighted with the response function $V(\lambda)$ related to the light-adapted eye (see equation [1]), they can be used for certain radiation protection purposes associated with specific optical sources. More details are to be found in Sl80. Limits of exposure for the skin and eye to UV radiation can be expressed in terms of the effective irradiance whereby the irradiance of a broadband source is weighted by a spectral effectiveness function S_λ (Chapters 3 and 4).

Levels of exposure to *ultrasound* are specified in terms of acoustic intensity (W/m^2) at the point of interest in the living organism and by the exposure duration, although acoustic pressure may prove to be more useful for specifying exposure limits to short-pulse ultrasound. For practical purposes of exposure control, the free-field time-averaged acoustic intensity in water and the duration of exposure are often measured and stated.

Exposure limits for *airborne ultrasound* are specified in terms of "acoustic pressure level" $L_p = 20 \log_{10}(p/p_0)$, where p_0 is the root mean square acoustic pressure (20 μPa), equivalent to an acoustic intensity $I_0 = 10^{-12}$ W/m^2 in air (Chapter 7).

CONCLUDING REMARKS

This document was prepared to provide a comprehensive reference for concepts, quantities, and units used in the whole range of NIR protection.

The physical phenomena of the NIRs belong to the classical fields of electromagnetism, optics, and acoustics. A well-developed and internationally accepted terminology exists for each of these fields. It can be utilized for the purposes of NIR protection, under consideration of the practical circumstances and biological constraints. Standardized names of the quantities, SI units, and the relevant definitions are compiled in Appendix 1.

The international organizations listed in the section on current standardization activities in the field of radiation, p. 9 have achieved a considerable degree of harmonization between closely related fields, especially between optics and other electromagnetic radiation. As Table 1 shows, this consists primarily in a harmonization of *radiometric* physical concepts, with some remaining differences in the names of the quantities. The same observation can be made by comparing these quantities with the quantities used for ionizing radiation. In optics the dualism between the radiometric and the photometric quantities still exists but is manageable due to a clear system of correspondence between the quantities of each kind (Table 2). Although ELF fields are describable in the terms of the electromagnetic theory, radiometric quantities are not relevant in this case since far field conditions do not occur in any practical case. The terminology for the ultrasound quantities is independent and self-consistent, and only incompletely harmonized with the quantities for other radiations (compare the terms "acoustic intensity" and "radiant intensity," see Table 1).

Interaction coefficients reflect to a higher degree the peculiarities of interaction mechanisms in different wavelength regions (Table 3). The existing nonuniformity is larger than in the case of radiometry, and more work should be done to harmonize the terminology.

From Table 4 it can be concluded that *exposure quantities* (dosimetric quantities) show a great diversity and nonuniformity. From the previous discussion it may be concluded that this diversity is probably partly due to divergent tradition. However, important factors contributing to the inherent differences are:

- physical circumstances prevailing at different wavelengths, and different wave propagation and interaction phenomena,
- appropriate measurement techniques, and
- physical factors determining biological effects according to differences between thermal, mechanical, and electric and magnetic action mechanisms.

It seems at the moment that the introduction of unified dosimetric concepts, such as specific absorption rate, is not practicable for *all* types of NIR.

In conclusion, coexistence of different quantities for different frequency ranges and purposes appears as a necessity, both for intrinsic and practical reasons, at least at the present time. No attempt was therefore made at this time to achieve harmonization between all subspecialties in this field. The committee is of the opinion that existing possibilities of introducing a greater uniformity, both from the conceptual point of view and from the standpoint of practical radiation protection, require further serious examination.

REFERENCES

AIUM81 American Institute of Ultrasound in Medicine/National Electrical Manufacturers Association, 1981, *Safety Standard for Diagnostic Ultrasound Equipment,* (Draft V, January 27) (Oklahoma City, OK: AIUM).

BIPM73 Bureau International des Poids et Mesures, 1973, *Le Système International d'Unités (SI)* (Sèvres: BIPM). Translated into English as *SI, The International System of Units* (National Physical Laboratory, Her Majesty's Stationery Office, London, 1977).

CIE70 International Commission on Illumination/International Electrotechnical Commission, 1970, *International Lighting Vocabulary* (3rd Edn) (Paris: Bureau Central de la C.I.E.).

CIP54 Comité International de Photobiologie, 1954, *Résolution* (Amsterdam: CIP). (Now available from Association Internationale de Photobiology, Secretariat General, Department of Plant Physiology, Fack, S-220 07, Lund, Sweden).

ICRU80 International Commission on Radiation Units and Measurements, 1980, "Radiation Quantities and Units," *ICRU Report* 33 (Bethesda, MD: ICRU).

ISO82 International Organization for Standardization, 1982, "Units of measurement," *ISO Standards Handbook* 2 (2nd Edn), (Geneva: ISO Central Secretariat).

ITU81 International Telecommunication Union, 1981, *Radio Regulations,* Revision 1981 (Geneva: ITU).

IUPAP78 International Union of Pure and Applied Physics, 1978, *IUPAP Symbols, Units and Nomenclature in Physics,* Document U.I.P. 20 (Quebec, Canada: IUPAP, University Laval).

Jo45 Joos G., 1945, *Lehrbuch der Theoretischen Physik,* Leipzig.

Mo79 Mohr H. and Schäfer E., 1979, "Uniform terminology for radiation: a critical comment," *Photochem. Photobiol.* **29,** 1061–1062.

NCRP81 National Council on Radiation Protection and Measurements, 1981, "Radiofrequency Electromagnetic Fields. Properties, Quantities and Units, Biophysical Interaction, and Measurements," *NCRP Report No.* 67 (Washington, DC: NCRP).

Ph84 Pergamon Press Ltd., 1984, "Notes for Contributors," *Photochem. Photobiol.* **39** (January).

Ru77 Rupert C. S. and Latarjet R., 1977, "Toward a nomenclature and dosimetric scheme applicable to all radiations," *Photochem. Photobiol.* **28,** 3–5.

Sl80 Sliney D. and Wolbarsht M., 1980, *Safety with Lasers and Other Optical Sources,* pp. 693–761 (New York and London: Plenum Press).

UN81 United Nations Environment Programme/World Health Organization/International Radiation Protection Association, 1981, "Radiofrequency and microwaves," *Environmental Health Criteria No.* 16 (Geneva: WHO).

Yo75 Youmans H. D. and Ho H. S., 1975, "Development of dosimetry for RF and microwave radiation. I. Dosimetric quantities for RF and microwave electromagnetic fields," *Health Phys.* **29,** 313–316.

APPENDIX 1: DEFINITIONS AND ABBREVIATIONS OF QUANTITIES AND UNITS

(The tables corresponding to Appendix 1 can be found on pages 24–39.)

Table 1: Periodic and radiation phenomena

This table contains those quantities and units that are common to all periodic and radiation phenomena. Table 1 can therefore be regarded as complementing the three following sections.

Table 2: Quantities and units of electromagnetic radiation and fields

For electricity and magnetism, different systems of equations have been developed, depending on the number and the particular choice of base quantities on which the system of equations is founded. For the purpose of this document, only the system of equations with four base quantities and with the four base units metre, kilogram, second, and ampere (SI units) is used.

Table 2 refers to electromagnetic radiation in general with emphasis on radiofrequency radiation; the special optical quantities and units are contained in Table 3. The table is introduced with the quantities and units describing electric conduction phenomena, both stationary and time-dependent, and with the terms used to quantify electric and magnetic fields including ELF.

Spectral distributions and spectral quantities can be defined in the same way as shown in Table 3 for optical radiation.

The term "factor" is used here to designate an amplitude ratio and is dimensionless (see 2.39); the term "coefficient" is used in the context of the exponential attenuation law and has the dimension length^{-1} (see 2.40). The corresponding terminology is used in Tables 3 and 4.

Table 3: Quantities and units of optical radiation

This table contains the radiometric quantities and units (3.1–3.19), the photometric quantities and units (3.20–3.27), the factors by which these are interrelated (3.28–3.30), and the material-dependent coefficients (3.31–3.38).

In several definitions, the same symbol is used for a pair of corresponding radiometric and photometric quantities with the understanding that subscripts e for radiometric (energetic) and v for photometric (visual) will be added whenever confusion between these quantities might otherwise occur.

The adjective "spectral" is used to designate (a) wavelength distribution functions, for example, $P_\lambda = \mathrm{d}P/\mathrm{d}\lambda$, the spectral radiant power (3.4), and (b) quantities that are functions of wavelength, for example, $\alpha(\lambda)$, the spectral absorptance (3.31). Note that P and P_λ have different dimensions.

The table does not contain the terms used to describe light in the particle description since these quantities are presently not being considered for use in radiation protection.

Table 4: Ultrasound

The compilation of acoustic quantities and units refers primarily to ultrasound, that is, to acoustic pressure waves with frequencies beyond 20 kHz. All of these quantities, however, are also applicable to audible sound. The term loudness level (4.18) only applies to the audible region of frequencies, but has been incorporated for comparison with the term acoustic pressure level. Power levels and pressure levels are expressed in decibels; the term Neper has not been included for the sake of simplicity (1 Np ≙ 8.686 dB).

The intensity terms collected in items 4.11–4.15 may be regarded as an example for a set of definitions useful to characterize a radiation field with spatial and temporal modulation (interference pattern, focusing, pulse mode). These terms could be adapted to other forms of NIR if necessary.

APPENDIX 2: PREFIXES INDICATING DECIMAL MULTIPLES OR SUBMULTIPLES OF UNITS

Multiple	Name of Prefix	Symbol
10^{18}	exa	E
10^{15}	peta	*P*
10^{12}	tera	T
10^{9}	giga	G
10^{6}	mega	M
10^{3}	kilo	k
10^{2}	hecto	h
10	deca	da

Multiple	Name of Prefix	Symbol
10^{-1}	deci	d
10^{-2}	centi	c
10^{-3}	milli	m
10^{-6}	micro	μ
10^{-9}	nano	n
10^{-12}	pico	p
10^{-15}	femto	f
10^{-18}	atto	a

Symbols for prefixes should be printed in roman (upright) type without space between the prefix and the symbol for the unit.

Compound prefixes should not be used.
Example: Write nm (nanometre) instead of mμm.

APPENDIX 3: ALPHABETIC INDEX OF QUANTITIES AND UNITS

The numbers given in this index refer to the classification of quantities and units in Appendix 1.

(Appendix 3 continues on pg. 40)

TABLE 1. *Periodic and radiation phenomena*

Item No.	Quantity	Symbol	Definition	Remarks
1.1	angle (plane angle)	α, φ	The angle between two half-lines terminating at the same point is defined as the ratio of the arc cut out on a circle (with its centre at that point) to the radius of the circle	According to this definition, the angle is a dimensionless quantity. For the unit of the angle, it is convenient to use the special name radian instead of the number 1
1.2	solid angle	Ω	The solid angle of a cone is defined as the ratio of the area cut out on a spherical surface (with its centre at the apex of that cone) to the square of the radius of the sphere	According to this definition, the solid angle is a dimensionless quantity. For the unit of the solid angle, it is convenient to use the special name steradian instead of the number 1
1.3	period, periodic time	T	Time of one cycle	
1.4	time constant of an exponentially varying quantity	τ	Time after which the quantity would reach its limit if it maintained its initial rate of variation	If a quantity is a function of time given by $F(t) = A + Be^{-t/\tau}$ then τ is the time constant
1.5	duty factor *	d	Ratio between the pulse duration, t_{pulse}, and the pulse period, T of a pulse-train, $d = t_{pulse}/T$	This quantity is dimensionless
1.6	frequency	f, ν	$f = 1/T$	
1.7	angular frequency	ω	$\omega = 2\pi f$	
1.8	bandwidth *	Δf	Difference $f_1 - f_2$ of the frequencies f_1 and f_2 at which the amplitude spectrum is 50 % (- 6 dB) of its maximum value	
1.9	center frequency *	f_c	Arithmetic mean $\frac{f_1 + f_2}{2}$ of the frequencies f_1 and f_2 defining bandwidth	
1.10	frequency interval		Binary logarithm of the ratio between the higher and the lower frequency, $\lg_2(f_2/f_1)$	This quantity is dimensionless
1.11	wavelength	λ		
1.12	wave number circular wave number	σ k	$\sigma = 1/\lambda$ $k = 2\pi\sigma$	The vector quantities corresponding to wave number and circular wave number are called wave vector and propagation vector respectively
1.13	amplitude level difference, field level difference	L_F	$L_F = 20 \lg(F_1/F_2)$ where F_1 and F_2 represent two amplitudes of the same kind	These quantites are dimensionless. If $P_1/P_2 = (F_1/F_2)^2$ then $L_P = L_F$. Similar names, symbols and definitions apply to level differences based on other quantities which are linear or quadratic functions of the amplitudes, respectively. The quantity on which the level difference is based should be specified in the name and the subscript of the symbol, e. g. field strength level difference L_E. A level difference from a standard situation is described simply as "level"
1.14	power level difference	L_P	$L_P = 10 \lg (P_1/P_2)$ where P_1 and P_2 represent two powers	
1.15	damping coefficient	δ	If a quantity is a function of time given by $f(t) = Ae^{-\delta t}\sin[\omega (t - t_0)]$ then δ is the damping coefficient	$\tau = 1/\delta$ is the time constant (relaxation time) of the amplitude. The quantity $\omega(t - t_0)$ is called the phase. For any two periodic functions with phases $\omega(t - t_1)$ and $\omega(t - t_2)$, the phase difference is $\omega(t_1 - t_2)$
1.16	logarithmic decrement	Λ	Product of damping coefficient and period	This quantity is dimensionless

* Not in ISO82.

TABLE 1. *Periodic and radiation phenomena (cont.)*

Item No.	Name of unit	International symbol for unit	Definition	Conversion factors	Remarks
1.1	radian	rad	1 rad is the angle between two radii of a circle which cut off on the circumference an arc equal in length to the radius		
	degree	°	$1° = \frac{\pi}{180}$ rad	1° = 0.017 453 3 rad	
1.2	steradian	sr	1 sr is the solid angle which, having its vertex in the centre of a sphere, cuts off an area of the surface of the sphere equal to that of a square with sides of length equal to the radius of the sphere		For a circular cone of solid angle 1 sr, the angle cut out in a plane that contains the axis of the cone is approximately 65.541° or 1.1439 rad
1.3	second	s	The second is the duration of 9 192 631 770 periods of the radiation corresponding to the transition between the two hyperfine levels of the ground state of the caesium-133 atom		
1.4	second	s			
1.5	1				
1.6	hertz	Hz	1 Hz is the frequency of a periodic phenomenon of which the period is 1 s 1 Hz = 1 s^{-1}		
1.7	reciprocal second	s^{-1}			
1.8	hertz	Hz			
1.9	hertz	Hz			
1.10	octave		The frequency interval between f_1 and f_2 is one octave if $f_2/f_1 = 2$		
1.11	metre	m	The metre is the distance traveled by light in a vacuum during 1/299 772 458 of a second	1 Å = 10^{-10}m (exactly) = 0.1 nm	
1.12	reciprocal metre, metre to the power minus one	m^{-1}			
1.13	decibel	dB	1 dB is the amplitude level difference when $20 \lg(F_1/F_2) = 1$		The numerical value of L_F expressed in decibels is given by $20 \lg(F_1/F_2)$, and the numerical value of L_P expressed in decibels is given by $10 \lg(P_1/P_2)$
1.14	decibel	dB	1 dB is the power level difference when $10 \lg(P_1/P_2) = 1$		
1.15	reciprocal second	s^{-1}			
1.16	1				

TABLE 2. *Electromagnetic radiation and fields*

Item No.	Quantity	Symbol	Definition	Remarks
2.1	electric current	I	Base quantity	
2.2	electric charge, quantity of electricity	Q	Integral of electric current over time	
2.3	surface density of charge	σ	Charge divided by surface area	
2.4	volume density of charge	ρ	Charge divided by volume	
2.5	electric field strength	$\underline{E}$	Force, exerted by electric field on an electric point charge, divided by the electric charge	effective electric field strength see the section on dosimetric quantities, p. 17
2.6	electric potential	φ	For electrostatic fields, a scalar quantity, the gradient of which, with reversed sign, is equal to the electric field strength	
	potential difference, tension, voltage	U	The potential difference between point 1 and point 2 is the line integral from 1 to 2 of the electric field strength $U = \varphi_1 - \varphi_2 = \int_1^2 E_s\, ds$	
2.7	electric flux density, displacement	$\underline{D}$	The electric flux density is a vector quantity, the divergence of which is equal to the volume density of charge	
2.8	capacitance	C	Charge divided by potential difference	
2.9	permittivity	ε	Electric flux density divided by electric field strength	
	permittivity of vacuum, electric constant	ε_0		$\varepsilon_0 = 1/(\mu_0 c_0^2)$ $= (8.854\,187\,818 \pm 0.000\,000\,071) \times 10^{-12}$ F/m
2.10	relative permittivity	ε_r	$\varepsilon_r = \varepsilon/\varepsilon_0$	This quantity is dimensionless
2.11	electric susceptibility	χ_e	$\chi_e = \varepsilon_r - 1$	This quantity is dimensionless
2.12	electric polarization	$\underline{P}$	$\underline{P} = \underline{D} - \varepsilon_0 \underline{E}$	
2.13	electric dipole moment	$\underline{p}$	The electric dipole moment is a vector quantity, the vector product of which with the electric field strength is equal to the torque	
2.14	current density	$\underline{J}$	A vector quantity the integral of which over a given surface is equal to the current flowing through that surface	$\underline{j}$ is also used
2.15	magnetic field strength	$\underline{H}$	The magnetic field strength is an axial vector quantity, the curl (rotation) of which is equal to the current density, including the displacement current	effective magnetic field strength see the section on dosimetric quantities, p. 17
2.16	magnetic flux density, magnetic induction	$\underline{B}$	The magnetic flux density is an axial vector quantity such that the force exerted on an element of current is equal to the vector product of this element and magnetic flux density	
2.17	magnetic flux	Φ	The magnetic flux across a surface element is the scalar product of the surface element and the magnetic flux density	
2.18	self inductance	L	For a conducting loop, the magnetic flux through the loop, caused by the current in the loop, divided by this current	
	mutual inductance	L_{12}	For two conducting loops, the magnetic flux through one loop, due to the current in the other loop, divided by this current	

TABLE 2. *Electromagnetic radiation and fields (cont.)*

Item No.	Name of unit	International symbol for unit	Definition	Conversion factors	Remarks
2.1	ampere	A	The ampere is that constant electric current which, if maintained in two straight parallel conductors of infinite length, of negligible circular cross-section, and placed 1 metre apart in vacuum, would produce between these conductors a force equal to 2×10^{-7} newton per metre of length		
2.2	coulomb	C	$1\ C = 1\ A \cdot s$		
2.3	coulomb per square metre	C/m^2			
2.4	coulomb per cubic metre	C/m^3			
2.5	volt per metre	V/m	$1\ V/m = 1\ N/C$		$1\ N = 1\ kg \cdot m \cdot s^{-2}$
2.6	volt	V	$1\ V = 1\ W/A$		
2.7	coulomb per square metre	C/m^2			
2.8	farad	F	$1\ F = 1\ C/V$		
2.9	farad per metre	F/m			
2.10					
2.11					
2.12	coulomb per square metre	C/m^2			
2.13	coulomb metre	$C \cdot m$			
2.14	ampere per square metre	A/m^2			
2.15	ampere per metre	A/m		$1\ Oe \cong 79.577\ A/m$	
2.16	tesla	T	$1\ T = 1\ N/(A \cdot m) = 1\ Wb/m^2 = 1\ V \cdot s/m^2$	$1\ G \cong 10^{-4}\ T$	
2.17	weber	Wb	$1\ Wb = 1\ V \cdot s$		
2.18	henry	H	$1\ H = 1\ Wb/A = 1\ V \cdot s/A$		

TABLE 2. *Electromagnetic radiation and fields (cont.)*

Item No.	Quantity	Symbol	Definition	Remarks
2.19	permeability	μ	Magnetic flux density divided by magnetic field strength	
	permeability of vacuum, magnetic constant	μ_0		$\mu_0 = 4\pi \cdot 10^{-7}$ H/m $= 12.566\ 370\ 614\ 4 \times 10^{-7}$ H/m
2.20	relative permeability	μ_r	$\mu_r = \mu/\mu_0$	This quantity is dimensionless
2.21	magnetic susceptibility	χ_m	$\chi_m = \mu_r - 1$	This quantity is dimensionless
2.22	electromagnetic moment, magnetic dipole moment	$\underline{m}$	The electromagnetic moment is a vector quantity, the vector product of which with the magnetic flux density is equal to the torque	
2.23	magnetization	$\underline{M}$	$\underline{M} = (\underline{B} / \mu_0) - \underline{H}$	
2.24	resistance (to direct current)	R	Electric potential difference divided by current when there is no electromotive force in the conductor	
2.25	conductance (to direct current)	G	$G = 1/R$	
2.26	resistivity	ρ	Electric field strength divided by current density when there is no electromotive force in the conductor	
2.27	conductivity	γ	$\gamma = 1/\rho$	
2.28	phase difference, phase displacement (for alternating current)	φ	When $U(t) = U_m \cos \omega t$ and $I(t) = I_m \cos(\omega t - \varphi)$ then φ is the phase displacement	This quantity is dimensionless
2.29	impedance, (complex impedance)	Z	The complex representation of potential difference divided by the complex representation of current	$Z = \lvert Z \rvert e^{j\varphi} = R + jX$
	modulus of impedance (impedance)	$\lvert Z \rvert$		$\lvert Z \rvert = \sqrt{R^2 + X^2}$
	reactance	X	Imaginary part of impedance	$X = L\omega - \frac{1}{C\omega}$ for a coil and a condenser in series
	resistance	R	Real part of impedance	
2.30	power (of an alternating current)	P	Product of time-dependent current and voltage	When $U(t) = U_m \cos \omega t$ and $I(t) = I_m \cos(\omega t - \varphi)$ then IU is the instantaneous power; $\frac{I_m U_m}{2}$ is the apparent power; $\frac{I_m U_m}{2} \cos \varphi$ is the active power
2.31	velocity of propagation of electromagnetic waves in vacuum *	c_0		$c_0 = \frac{1}{\sqrt{\varepsilon_0 \mu_0}} = (2.997\ 924\ 58 \pm 0.000\ 000\ 012) \times 10^8$ m/s
	velocity of propagation of electromagnetic waves in any medium	c		
2.32	radiant energy	Q, W	Energy emitted, transferred or received in the form of electromagnetic radiation	
2.33	radiant power, radiant energy flux	P, Φ	Power emitted, transferred or received in the form of electromagnetic radiation	

* ISO82 and IUPAP78 recommend symbol c for velocity of propagation of electromagnetic waves in vacuum.

TABLE 2. *Electromagnetic radiation and fields (cont.)*

Item No.	Name of unit	International symbol for unit	Definition	Conversion factors	Remarks
2.19	henry per metre	H/m	1 H/m = 1 Wb/(A · m) = 1 V · s/A · m)		
2.20					
2.21					
2.22	ampere metre squared	$A \cdot m^2$			The quantity magnetic dipole moment has the unit Wb · m
2.23	ampere per metre	A/m			
2.24	ohm	Ω	1 Ω = 1 V/A		
2.25	siemens	S	1 S = 1 A/V		$1\ S = 1\ \Omega^{-1}$
2.26	ohm metre	Ω · m			
2.27	siemens per metre	S/m			
2.28	radian	rad			Unit radian: see item 1.1
2.29	ohm	Ω	1 Ω = 1 V/A		
2.30	watt	W	1 W = 1 J/s = 1 V · A		In electric power technology, active power is expressed in watts (W) and apparent power in volt-amperes (V · A)
2.31	metre per second	m/s			
2.32	joule	J	1 J = 1 V · A · s		
2.33	watt	W			

TABLE 2. *Electromagnetic radiation and fields (cont.)*

Item No.	Quantity	Symbol	Definition	Remarks
2.34	electromagnetic energy density	w	In an electromagnetic field, the scalar products $\underline{D}\underline{E}$ and $\underline{B}\underline{H}$ determine energy density $w = \frac{\underline{D}\underline{E}}{2} + \frac{\underline{B}\underline{H}}{2}$	
2.35	Poynting vector	$\underline{S}$	The Poynting vector is equal to the vector product of electric field strength and magnetic field strength	
2.36	power (surface) density * energy flux density *	ψ	Radiant power incident on a small sphere, divided by the cross-sectional area of that sphere	For a plane wave, radiant power density is the time-dependent magnitude $\lvert\underline{S}\rvert$ of the Poynting vector. Its time average is $\lvert\overline{\underline{S}}\rvert$. For ionizing radiation, this quantity is called the energy fluence rate. In optics, this quantity is called "radiant flux density" or "spherical irradiance", see 3.5
2.37	power per solid angle****	I	Radiant power of electromagnetic radiation leaving a source in an element of solid angle at a given direction, divided by that element of solid angle. $I = dP/d\Omega$	In optics, this quantity is called "radiant intensity", see 3.9
2.38	Wave impedance (of an electromagnetic wave) *	Z	Quotient of the electric field strength and the corresponding magnetic field strength of a propagating wave at a point, $Z = \lvert\underline{E}\rvert / \lvert\underline{H}\rvert$	In general, Z depends on the phase difference between $\lvert\underline{E}\rvert$ and $\lvert\underline{H}\rvert$. For a plane wave in a non-dissipative isotropic medium the impedance has the value $Z_0 = \sqrt{\mu/\varepsilon}$ (intrinsic impedance). In a vacuum, $Z_0 = \sqrt{\mu_0/e_0} = 377\ \Omega$
2.39	reflection factor **	Γ	Ratio of the (electric or magnetic) field strength $\lvert\underline{F_r}\rvert$ of the reflected electromagnetic wave to the corresponding field strength $\lvert\underline{F_i}\rvert$ of the incident wave at a boundary between two media. $\Gamma = \lvert\underline{F_r}\rvert / \lvert\underline{F_i}\rvert$	In general, the reflection factor and the transmission factor depend on the phase difference between $\lvert\underline{F_t}\rvert$ and $\lvert\underline{F_i}\rvert$. For normal incidence, $\Gamma = (Z_2 - Z_1) / (Z_2 + Z_1)$ $\tau = 2Z_2 / (Z_2 + Z_1)$
	transmission factor **	τ	Ratio of the (electric or magnetic) field strength $\lvert\underline{F_t}\rvert$ of the transmitted electromagnetic wave to the corresponding field strength $\lvert\underline{F_i}\rvert$ of the incident wave at a boundary between two media. $\tau = \lvert\underline{F_t}\rvert / \lvert\underline{F_i}\rvert$	with $\lvert\tau\rvert - \lvert\Gamma\rvert = 1$. The power ratios for reflection and transmission *** are $\lvert\Gamma\rvert^2$ and $\lvert\tau\rvert^2$ respectively. These quantities are dimensionless.
2.40	power attenuation coefficient *	α	For an exponential change of the power density ψ of an electromagnetic wave with distance x, $\psi = \psi_0\, e^{-\alpha x}$ the exponent is the product of the attenuation coefficient and the distance. The contributions to attenuation by energy dissipation processes (absorption) and by wave scattering are expressed by $e^{-\alpha x} = e^{-\alpha_a x}\, e^{-\alpha_s x}$	The quantity $1/\alpha$ is denoted as attenuation length (depth of penetration). The amplitude in an electromagnetic wave is attenuated with distance x according to the "amplitude attenuation coefficient" $\alpha/2$ $\alpha = \alpha_a + \alpha_s$ Compare spectral linear attenuation coefficient (3.35) and spectral linear absorption coefficient (3.36) in optics.
2.41	specific absorption *	SA	Quotient of the incremental energy dW absorbed (dissipated) in matter of the incremental mass dm contained in a volume element dV of given density, and this mass dm $SA = dW/dm = dW/\rho dV$	For ionizing radiations, this quantity is called absorbed dose. The quantity dW/dV has been called "absorption density".
2.42	specific absorption rate *	SAR	Time derivative of the specitic absorption $SAR = \frac{d}{dt}(SA) = \frac{d}{dt}\left(\frac{dW}{dm}\right) = \frac{d}{dt}\left(\frac{dW}{\rho dV}\right)$	For ionizing radiations, this quantity is called absorbed dose rate. The quantity d/dt (dW/dV) has been called "absorption rate density".

* Not in ISO 31/5-1979 (ISO82). Adopted from (NCRP81).

** Not in ISO 31/5-1979. Adopted from G. Joos, Lehrbuch der theoretischen Physik, Leipzig 1945 (Jo45). In (NCRP81) these quantities are called reflection coefficient and transmission coefficient.

*** In photometry, the corresponding ratios are called spectral reflectance and spectral transmittance (see 3.32 and 3.33).

**** Not in ISO 31/5-1979 (ISO82).

TABLE 2. *Electromagnetic radiation and fields (cont.)*

Item No.	Name of unit	International symbol for unit	Definition	Conversion factors	Remarks
2.34	joule per cubic metre	J/m^3			
2.35	watt per square metre	W/m^2			
2.36	watt per square metre	W/m^2			
2.37	watt per steradian	W/sr			Unit steradian: see item 1.2
2.38	ohm	Ω			
2.39					
2.40	reciprocal metre	m^{-1}			1 db is the power level attenuation when $10\ lg(\psi_0/\psi) = 1$
2.41	joule per kilogram	J/kg			
2.42	watt per kilogram	W/kg			

TABLE 3. *Optical radiation*

Item No.	Quantity	Symbol	Definition	Remarks
3.1	radiant energy	Q, W	Energy emitted, transferred or received as optical radiation	$Q = \int Q_\lambda d\lambda$
3.2	spectral radiant energy	Q_λ, W_λ	The radiant energy in an infinitesimal wavelength interval divided by the range of that interval	$Q_\lambda = dQ/d\lambda$
3.3	radiant power, radiant energy flux	P, Φ	Power emitted, transferred or received as optical radiation	$P = \int P_\lambda d\lambda$
3.4	spectral radiant power	P_λ, Φ_λ	The radiant power in an infinitesimal wavelength interval divided by the range of that interval	$P_\lambda = dP/d\lambda$
3.5	radiant energy fluence rate, spherical irradiance *	φ	At a given point in space, the radiant power incident on a small sphere, divided by the cross-sectional area of that sphere	$\varphi = \int \varphi_\lambda d\lambda$ In an isotropic homogeneous radiation field, φ/c is the energy density, and the irradiance of a surface is $\varphi/4$. Radiant flux density is the integral of radiance over all directions: $\varphi = \int_{4\pi} L\, d\Omega$ (see 3.11)
3.6	spectral radiant energy fluence rate	φ_λ	The radiant energy fluence rate in an infinitesimal wavelength interval divided by the range of that interval	
3.7	radiant energy density	w	Radiant energy in an element of volume divided by that element	$w = \int w_\lambda d\lambda$
3.8	spectral radiant energy density	w_λ	The radiant energy density in an infinitesimal wavelength interval divided by the range of that interval	
3.9	radiant intensity	I	For a source in a given direction, the radiant power leaving the source, or an element of the source, in an element of solid angle containing the given direction, divided by that element of solid angle	$I = \int I_\lambda d\lambda$
3.10	spectral radiant intensity	I_λ	The radiant intensity in an infinitesimal wavelength interval divided by the range of that interval	
3.11	radiance	L	At a point of a surface and in a given direction, the radiant intensity dI at an element dA of the surface, divided by the area da of the orthogonal projection of this element on a plane perpendicular to the given direction: $L = \frac{dI}{da}$	$L = \int L_\lambda d\lambda$ At a light emitting surface, $da = dA \cdot \cos \Theta_s$ and $L = dI / (dA \cdot \cos \Theta_s)$ where Θ_s is the angle between the direction of radiation and the normal on the source surface. At a receptor surface, $da = dA \cdot \cos \Theta_r$ and $L = dI / (dA \cdot \cos \Theta_r)$ where Θ_r is the angle between the direction of radiation and the normal on the receptor surface.
3.12	time-integrated radiance **	K	Time integral of the radiance over a given time interval	$K = \int L\, dt$
3.13	spectral radiance	L_λ	The radiance in an infinitesimal wavelength interval divided by the range of that interval	$L_\lambda = dL/d\lambda$
3.14	radiant exitance	M	At a point of a surface, the radiant power leaving an element of the surface, divided by the area of that element	$M = \int M_\lambda d\lambda$
3.15	spectral radiant exitance	M_λ	The radiant exitance in an infinitesimal wavelength interval divided by the range of that interval	$M_\lambda = dM/d\lambda$
3.16	irradiance	E	At a point of a surface, the radiant power incident on an element of the surface, divided by the area of that element	$E = \int E_\lambda d\lambda$
3.17	spectral irradiance	E_λ	The irradiance in an infinitesimal wavelength interval divided by the range of that interval	$E_\lambda = dE/d\lambda$
3.18	radiant exposure*	H	Time integral of the irradiance	$H = \int H_\lambda d\lambda$
3.19	spectral radiant exposure	H_λ	The radiant exposure in an infinitesimal wavelength interval divided by the range of that interval	$H_\lambda = dH/d\lambda$

* (CIE 70).

** Not in ISO 31/6, 1980 (ISO 82).

TABLE 3. *Optical radiation (cont.)*

Item No.	Name of unit	International symbol for unit	Definition	Conversion factors	Remarks
3.1	joule	J		1 erg = 10^{-7}J	
3.2	joule per nanometre	J/nm		1 erg/Å = 10^{-6}J/nm	
3.3	watt	W			
3.4	watt per nanometre	W/nm			
3.5	watt per square metre	W/m^2			
3.6	watt per square metre nanometre	$W/(m^2 \cdot nm)$			
3.7	joule per cubic metre	J/m^3			
3.8	joule per cubic metre nanometre	$J/(m^3 \cdot nm)$			
3.9	watt per steradian	W/sr			Unit steradian: see item 1.2
3.10	watt per steradian nanometre	$W/(sr \cdot nm)$			
3.11	watt per square metre steradian	$W/(m^2 \cdot sr)$			
3.12	joule per square metre steradian	$J/(m^2 \cdot sr)$			
3.13	watt per square metre steradian nanometre	$W/(m^2 \cdot sr \cdot nm)$			
3.14	watt per square metre	W/m^2			
3.15	watt per square metre nanometre	$W/(m^2 \cdot nm)$			
3.16	watt per square metre	W/m^2			
3.17	watt per square metre nanometre	$W/(m^2 \cdot nm)$			
3.18	joule per square metre	J/m^2			
3.19	joule per square metre nanometre	$J/(m^2 \cdot nm)$			

TABLE 3. *Optical radiation (cont.)*

Item No.	Quantity	Symbol	Definition	Remarks
3.20	emissivity	ε	Ratio of radiant exitance of a thermal radiator to that of an ideal black body radiator at the same temperature	
3.21	luminous intensity	I	Basic quantity	
3.22	luminous flux	Φ	The luminous flux $d\Phi$ of a source of luminous intensity I in an element of solid angle $d\Omega$ is $d\Phi = I\ d\Omega$	$\Phi = \int \Phi_\lambda d\lambda$
3.23	spectral luminous flux	Φ_λ	The luminous flux in an infinitesimal wavelength interval divided by the range of that interval	$\Phi_\lambda = d\Phi/d\lambda$
3.24	quantity of light	Q	Time integral of luminous flux	
3.25	luminance	L	At a point of a surface and in a given direction, the luminous intensity of an element of the surface, divided by the area of the orthogonal projection of this element on a plane perpendicular to the given direction	
3.26	luminous exitance	M	At a point of a surface, the luminous flux leaving an element of the surface, divided by the area of that element.	
3.27	illuminance	E	At a point of a surface, the luminous flux incident on an element of the surface, divided by the area of that element	
3.28	light exposure	H	Time integral of illuminance	
3.29	spectral luminous efficiency	$V(\lambda)$	Relative sensitivity of the light-adapted eye at wavelength λ	Standard values of $V(\lambda)$ relating to the light-adapted eye were adopted by the International Commission on Illumination in 1924 and approved by the Comité International des Poids et Mesures in 1933 $V(\lambda) \equiv 1$ for λ = 555 nm
3.30	maximum spectral luminous efficacy	K_m	Factor converting spectral radiant flux into spectral luminous flux at λ = 555 nm. The value K_m = 683 lm/W follows from the definition of the SI-unit candela (see 3.20)	For the normal light-adapted eye under certain standardized conditions K_m = K(555 nm) = 683 lm/W. Relationship between radiometric and photometric quantities: $\Phi_{v\lambda} = K_m V(\lambda)\Phi_{e\lambda}$
3.31	spectral luminous efficacy	$K(\lambda)$	$K(\lambda) = K_m V(\lambda)$	$\Phi_{v\lambda} = K(\lambda)\Phi_{e\lambda}$
3.32	spectral absorptance	$\alpha(\lambda)$	Ratio of the spectral radiant or luminous flux absorbed to that of the incident radiation.	These quantities are dimensionless. The quantities 3.30, 3.31 and 3.32 are also called spectral absorption factor, spectral reflection factor and spectral transmission factor respectively
3.33	spectral reflectance	$\rho(\lambda)$	Ratio of the spectral radiant or luminous flux reflected to that of the incident radiation	
3.34	spectral transmittance	$\tau(\lambda)$	Ratio of the spectral radiant or luminous flux transmitted to that of the incident radiation.	
3.35	spectral radiance factor	$\beta(\lambda)$	At a point of a surface and in a given direction the ratio of the spectral radiance of a non-self-radiating body to that of a perfect diffuser under identical irradiation conditions	
3.36	spectral linear attenuation coefficient, linear extinction coefficient	$\mu(\lambda)$	The relative decrease in spectral radiant or luminous flux of a collimated beam of electromagnetic radiation during traversal of an infinitesimal layer of a medium, divided by the thickness of that layer	μ/ρ, where ρ is the density of the medium, is called mass attenuation coefficient
3.37	spectral linear absorption coefficient	$a(\lambda)$	The part of the linear attenuation coefficient that is due to absorption	a/ρ, where ρ is the density of the medium, is called mass absorption coefficient
3.38	spectral molar absorption coefficient	$\kappa(\lambda)$	$\kappa(\lambda) = \frac{a(\lambda)}{c}$, where c is the molar concentration	
3.39	spectral refractive index	$n(\lambda)$	The ratio of the velocity of electromagnetic radiation in vacuo to the phase velocity of electromagnetic radiation of a specified frequency in a medium	This quantity is dimensionless

TABLE 3. *Optical radiation (cont.)*

Item No.	Name of unit	International symbol for unit	Definition	Conversion factors	Remarks
3.20					
3.21	candela	cd	Candela is the luminous intensity in a given direction of a source which emits monochromatic radiation of frequency $540 \cdot 10^{12}$ Hz and of which radiant intensity in that direction is 1/683 watt/sr		
3.22	lumen	lm	1 lm = 1 cd · sr		
3.23	lumen per nanometre	lm/nm			
3.24	lumen second	lm · s		1 lm · h = 3600 lm · s (exactly)	
3.25	candela per square metre	cd/m^2		1 sb = 1 cd/cm^2 (stilb)	
3.26	lumen per square metre	lm/m^2			
3.27	lux	lx			
3.28	lux second	lx · s		1 lx · h = 3600 lx · s (exactly)	
3.29					
3.30	lumen per watt	lm/W			
3.31	lumen per watt	lm/W			
3.32					
3.33					
3.34					
3.35					
3.36	reciprocal metre	m^{-1}			
3.37	reciprocal metre	m^{-1}			
3.38	square metre per mole	m^2/mol			
3.39					

TABLE 4. *Ultrasound*

Item No.	Quantity	Symbol	Definition	Remarks
4.1	velocity of sound	c	Velocity of an acoustic wave	For example, in a homogeneous liquid, the velocity of sound is $c = \sqrt{K_a/\rho}$ where K_a is the adiabatic bulk modulus and ρ the density
4.2	acoustic particle displacement	ξ	Instantaneous displacement of a particle of the medium from what would be its position in the absence of acoustic waves	In an ultrasound field, the quantities p, ξ, v, and a have instantaneous values varying in time. Their amplitudes are denoted by $\hat{p}$, $\hat{\xi}$, $\hat{v}$, and $\hat{a}$ respectively.
4.3	acoustic particle velocity	v	$v = \frac{\partial \xi}{\partial t}$	
4.4	acoustic particle acceleration	a	$a = \frac{\partial v}{\partial t}$	
4.5	static pressure	p_s	Pressure that would exist in the absence of acoustic waves	
4.6	acoustic pressure	p	The difference between the instantaneous total pressure and the static pressure	
4.7	acoustic radiation pressure *	p_{rad}	Temporal average of the pressure exerted on the boundary between two media	
4.8	acoustic energy density	w	Mean acoustic energy in a given volume divided by that volume	If the energy density is varying with time, the mean must be taken over an interval during which the sound may be considered statistically stationary. The acoustic energy density at a point in the far field region can be expressed by $w = \frac{1}{2}\rho\hat{v}^2 = \frac{1}{2}\frac{\hat{p}^2}{\rho c^2}$
4.9	acoustic power acoustic energy flux	P	Acoustic energy emitted or transferred in a certain time interval, divided by the duration of that interval	
4.10	acoustic intensity **	I	Acoustic power incident on a small sphere, divided by the cross-sectional area of that sphere	The acoustic intensity in a specified direction at a point is the acoustic power transmitted in the specified direction through an area normal to this direction at the point considered, divided by this area
4.11	spatial peak-temporal peak intensity **	I_{SPTP}	Temporal peak intensity at the point in the acoustic field where the temporal peak intensity is a maximum	
4.12	spatial peak-pulse average intensity **	I_{SPPA}	Pulse average intensity at the point in the acoustic field where the pulse average intensity is a maximum	
4.13	spatial peak-temporal average intensity **	I_{SPTA}	Temporal average intensity at the point in the acoustic field where the temporal average intensity is a maximum	
4.14	spatial average-pulse average intensity **	I_{SAPA}	Pulse average intensity averaged over the beam cross-sectional area	
4.15	spatial average-temporal average intensity **	I_{SATA}	Temporal average intensity averaged over the beam cross-sectional area	
4.16	acoustic pressure level	L_p, L	$L_p = 20 \lg(p/p_0)$ where p is the root-mean square acoustic pressure and p_0 a reference pressure	This quantity is dimensionless. The reference pressure must be explicitly stated, generally $p_0 = 20\ \mu Pa$ The subscript p is often omitted, when other subscripts are needed
4.17	acoustic power level	L_P	$L_P = 10 \lg(P/P_0)$ where P and P_0 are a given acoustic power and a reference power, respectively	This quantity is dimensionless. The reference power must be explicitly stated, generally $P_0 = 10^{-12}$ W

* Not in ISO 31/7-1978 (ISO82).
** Not in ISO 31/7-1978 (ISO82). Adopted from AIUM/NEMA standard (AIUM/NEMA 81).

TABLE 4. *Ultrasound (cont.)*

Item No.	Name of unit	International symbol for unit	Definition	Conversion factors	Remarks
4.1	metre per second	m/s			
4.2	metre	m			
4.3	metre per second	m/s			
4.4	metre per square second	m/s^2			
4.5	pascal	Pa		1 bar = 10^5 Pa (exactly)	
4.6	pascal	Pa		1 bar = 10^5 Pa (exactly)	
4.7	pascal	Pa			
4.8	joule per cubic metre	J/m^3			
4.9	watt	W			
4.10	watt per square metre	W/m^2			
4.11	watt per square metre	W/m^2			
4.12	watt per square metre	W/m^2			
4.13	watt per square metre	W/m^2			
4.14	watt per square metre	W/m^2			
4.15	watt per square metre	W/m^2			
4.16	decibel	dB	1 dB is the sound pressure level when $20\ lg(p/p_0) = 1$		The numerical value of L_p expressed in decibels is given by $20\ lg(p/p_0)$
4.17	decibel	dB	1 dB is the sound power level when $10\ lg(P/P_0) = 1$		The numerical value of L_P expressed in decibels is given by $10\ lg(P/P_0)$

TABLE 4. *Ultrasound (cont.)*

Item No.	Quantity	Symbol	Definition	Remarks
4.18	loudness level	L_N	$L_N = 20\ \lg(p_{eff}/p_0)_{1\ kHz}$ where p_{eff} is the effective (root-mean-square) sound pressure of a standard pure tone of 1 kHz which is judged by a normal observer under standardized listening conditions as being equally loud, and where p_0 = 20 µPa.	This quantity is dimensionless
4.19	specific acoustic impedance *	Z	At a point in a medium, the complex representation of acoustic pressure divided by the complex representation of particle velocity. $Z = p/v$	In general, Z depends on the phase difference between p and v. For a plane progressive wave in a non-dissipative medium the specific acoustic impedance has the value ρc. This is called the characteristic impedance.
4.20	pressure reflexion factor **	r	Ratio of the acoustic pressure of the wave reflected at a boundary to the pressure of the incident wave $r = p_r/p_i$	These quantities are dimensionless. For normal incidence on a boundary between media with characteristic impedances Z_1 and Z_2, $r = (Z_2 - Z_1)/(Z_2 + Z_1)$
4.21	pressure transmission factor **	t	Ratio of the acoustic pressure of the wave transmitted through a boundary to the pressure of the incident wave $t = p_t/p_i$	$t = 2Z_2/(Z_2 + Z_1)$
4.22	intensity reflection factor **	R	Ratio of the acoustic intensity reflected at a boundary to the incident intensity $R = I_r/I_i$	$R = (Z_2 - Z_1)^2/(Z_2 + Z_1)^2$
4.23	intensity transmission factor **	T	Ratio of the acoustic intensity transmitted through a boundary to the incident intensity $T = I_t/I_i$	$T = 4Z_2Z_1/(Z_2 + Z_1)^2$
4.24	amplitude attenuation coefficient	α	For an exponential change of the pressure amplitude A of an acoustic wave with distance x, $A = A_0\ e^{-\alpha x}$, the exponent is the product of the amplitude attenuation coefficient and the distance.	The quantity $1/\alpha$ is denoted as attenuation length. The energy density in an acoustic wave is attenuated with distance x according to the "energy attenuation coefficient" 2α
4.25	amplitude absorption coefficient **	α_a	The contributions to attenuation by absorption and scattering are expressed by	$\alpha = \alpha_a + \alpha_s$
	amplitude scattering coefficient **	α_s	$e^{-\alpha x} = e^{-(\alpha_a + \alpha_s)x}$	
4.26	level attenuation coefficient **	α'	For a linear change of acoustic pressure level or power level with distance x, $L(x) = L(0) - \alpha' x$, the slope of the curve is the level attenuation coefficient	
4.27	level absorption coefficient **	α'_a	For a linear change of acoustic pressure level or power level with distance x, $L(x) = L(0) - (\alpha'_a + \alpha'_s)x$,	$\alpha' = \alpha'_a + \alpha'_s$
	level scattering coefficient **	α'_s	the slope of the curve is the sum of the level absorption and scattering coefficient	
4.28	intensity backscattering coefficient **	B	Quotient $\Delta P_B/(\Delta V \cdot \Delta\Omega \cdot I_0)$, where ΔP_B is the acoustic power backscattered from volume ΔV into solid angle $\Delta\Omega$ at scattering angle 180°, and I_0 is the incident acoustic intensity	

* Denoted as "characteristic impedance" in ISO 31/7-1978 (ISO82).
** Not in ISO 31/7-1978 (ISO82).

TABLE 4. *Ultrasound (cont.)*

Item No.	Name of unit	International symbol for unit	Definition	Conversion factors	Remarks
4.18	phon		1 phon is the loudness level when $20 \lg(p_{eff}/p_o)_{1\ kHz} = 1$		The numerical value of L_N expressed in phons is given by $20 \lg(p_{eff}/p_o)_{1\ kHz}$
4.19	pascal second per metre	$\frac{Pa \cdot s}{m}$			
4.20					Numerical values in dB are obtained as 20 lg r
4.21					20 lg t
4.22					10 lg R
4.23					10 lg T
4.24	reciprocal metre	m^{-1}			
4.25	reciprocal metre	m^{-1}			
4.26	decibel per metre	dB/m		$\alpha' = (8.686\ dB) \cdot \alpha$	
4.27	decibel per metre	dB/m		$\alpha'_a = (8.686\ dB) \cdot \alpha_a$ $\alpha'_s = (8.686\ dB) \cdot \alpha_s$	
4.28	reciprocal metre per steradian	$m^{-1}\ sr^{-1}$			

APPENDIX 3 (cont.)

CHAPTER 3

GUIDELINES ON LIMITS OF EXPOSURE TO ULTRAVIOLET RADIATION OF WAVELENGTHS BETWEEN 180 nm AND 400 nm (INCOHERENT OPTICAL RADIATION)

A DOCUMENT entitled *Environmental Health Criteria 14, Ultraviolet Radiation* (UN79) was published in 1979 under the joint sponsorship of the United Nations Environment Programme (UNEP), the World Health Organization (WHO), and the International Radiation Protection Association (IRPA). The document contains a review of the biological effects reported from exposure to ultraviolet radiation (UVR) and serves as the scientific rationale for the development of these guidelines. The important publications that relate most directly to the guidelines (some of which have appeared since the Environmental Health Criteria document was drafted) are referenced in the rationale (Appendix).

The purpose of these guidelines is to deal with the basic principles of protection against noncoherent UVR, so that they may serve as guidance to the various international and national bodies or individual experts who are responsible for the development of regulations, recommendations, or codes of practice to protect the workers and the general public from the potentially adverse effects of UVR.

The IRPA/INIRC (International Non-Ionizing Radiation Committee) recognized that when standards on exposure limits are established, various value judgments are made. The validity of scientific reports has to be considered, and extrapolations from animal experiments to effects on humans have to be made. Cost vs. benefit analyses are necessary, including economic impact of controls. The limits in these guidelines were based on the scientific data and no consideration was given to economic impact or other nonscientific priorities. However, from presently available knowledge, the limits should provide a healthy working or living environment from exposure to UVR under all normal conditions.

The IRPA Associate Societies as well as a number of competent institutions and individual experts were consulted in the preparation of the guidelines and their cooperation is gratefully acknowledged.

The guidelines on limits of exposure to UVR were first approved by the IRPA Executive Council in May 1984 and published in *Health Physics* in August 1985. They were subsequently modified to account for more recent data in the near-ultraviolet UV-A spectral region (*Health Physics,* June 1989). During the preparation of the guidelines, the composition of the IRPA/INIRC (International Non-Ionizing Radiation Committee) was as follows:

H. P. Jammet, Chairman (France)
B. F. M. Bosnjakovic (Netherlands)
P. Czerski (Poland)
M. Faber (Denmark)

D. Harder (Germany)
J. Marshall (Great Britain)
M. H. Repacholi (Australia)
D. H. Sliney (U.S.A.)
J. C. Villforth (U.S.A.)
A. S. Duchêne, Scientific Secretary (France)

In addition, the following members participated in their later updating: J. Bernhardt (Germany), M. Grandolfo (Italy), B. Knave (Sweden), J. A. J. Stolwijk (U.S.A.).

INTRODUCTION

Ultraviolet radiation occupies that portion of the electromagnetic spectrum from 100 to 400 nm. In discussing UVR biological effects the International Commission on Illumination (CIE) has divided the UV spectrum into three bands. The band 315 to 380–400 nm is designated as UV-A, 280–315 nm as UV-B, and 100–280 nm as UV-C (CIE70). Wavelengths below 180 nm (vacuum UV) are of little practical biological significance since they are readily absorbed in air. Ultraviolet radiation is used in a wide variety of medical and industrial processes and for cosmetic purposes. These include photocuring of inks and plastics (UV-A and UV-B), photoresist processes (all UV), solar simulation (all UV), cosmetic tanning (UV-A and UV-B), fade testing (UV-A and UV-B), dermatology (all UV), and dentistry (UV-A). Even though the principal operating wavelengths for most of these processes are in the UV-A, almost always some shorter wavelength (UV-B and UV-C) radiation and violet light are emitted as well. Many industrial applications employ arc sources for heat or light (e.g., welding), which also produce unwanted UVR for which control measures may be necessary. While it is generally agreed that some low-level exposure to UVR benefits health (UN79), there are adverse effects that necessitate the development and use of exposure limits (EL) for UVR.

The most significant adverse health effects of exposure to UVR have been reported at wavelengths below 315 nm, known collectively as actinic ultraviolet. This guideline has been limited to wavelengths greater than 180 nm where UVR is transmitted through air. The most restrictive limits are for exposure to radiation having those wavelengths less than 315 nm.

PURPOSE AND SCOPE

The purpose of this document is to provide guidance on maximal limits of exposure to UVR in the spectral region between 180 nm and 400 nm and represent conditions under which it is expected that nearly all individuals may be repeatedly exposed without adverse effect (see section on Special Considerations). These EL values for exposure of the eye or the skin may be used to evaluate potentially hazardous exposure from UVR; for example, from arcs, gas and vapor discharges, fluorescent lamps, incandescent sources, and solar radiation. The limits do not apply to UV lasers. Most incoherent UV sources are broadband, although single emission lines can be produced from low-pressure gas discharges. These values should be used as guides in the control of exposure to both pulsed and continuous sources where the exposure duration is not less than 0.1 μs. These ELs are below levels that would be used for UV exposures of patients required as a part of medical treatment or for elective cosmetic purposes. These ELs are exceeded by noonday sunlight overhead at 0°–40° latitude within 5–10 minutes in the summertime. The ELs should be considered absolute limits for the eye, and "advisory" for the skin because of the wide range of susceptibility to skin injury depending on skin type. The ELs should be adequate to protect lightly pigmented individuals.

BASIC CONCEPTS

This document makes use of the spectral band designations of the CIE. Unless otherwise stated, UV-A is from 315 to 400 nm, UV-B is from 280 to 315 nm, and UV-C is from 100 to 280 nm (CIE70). It should be noted that some specialists follow this general scheme but take the dividing line between UV-A and UV-B at 320 nm. The UVR exposure should be quantified in terms of an irradiance E (W/m^2 or W/cm^2) for continuous exposure or in terms of a radiant exposure H (J/m^2 or J/cm^2) for time-limited (or pulsed) exposures of the eye and skin. The geometry of exposure to UVR is very important. For example, the eyes (and to a lesser extent the skin) are somewhat anatomically protected against UVR exposure from overhead sources (e.g., the sun overhead) (UN79). The limits should be applied to exposure directed perpendicular to those surfaces of the body facing the radiation source, measured with an instrument having cosine angular response (UN79). The irradiance and the radiant exposure should be averaged over the area of a circular measurement aperture not greater than 1 mm in diameter. These ELs should be used as guides in the control of exposure to UV sources and as such are intended as upper limits for nontherapeutic and non-elective exposure. The ELs should be considered as absolute limits for ocular exposure. The ELs were developed by considering lightly pigmented populations (i.e., Caucasian) with greatest sensitivity and genetic predisposition. Exposure during sun bathing and tanning under artificial sources may well exceed these limits but exposed individuals should be advised that some health risk is incurred from such activity. Eye protection is always required during therapeutic exposures. Nevertheless, occasional exposures to conditioned skin may not result in adverse effects. The rationale for the UVR exposure limits is provided in the Appendix.

EXPOSURE LIMITS

The EL for both general and occupational exposure to UVR incident on the skin or eye where irradiance values are known and the exposure duration is controlled are as follows:

For the near-ultraviolet UV-A spectral region (315–400 nm), the total radiant exposure incident on the unprotected eye should not exceed 1.0 J cm^{-2} (10 kJ m^{-2}) within an 8-hour period and the total 8-hour radiant exposure incident on the unprotected skin should not exceed the values given in Table 1. Values for the relative spectral effectiveness, S_λ, are given up to 400 nm to expand the action spectrum into the UV-A for determining the EL for skin exposure.

For the actinic UV spectral region (UV-C and UV-B from 180 to 315 nm), the radiant exposure incident upon the unprotected skin or eye within an 8-hour period should not exceed the values given in Table 1.

To determine the effective irradiance of a broadband source weighted against the peak of the spectral effectiveness curve (270 nm), the following weighting formula should be used:

$$E_{eff} = \sum E_\lambda \cdot S_\lambda \cdot \Delta_\lambda$$

where:

E_{eff} = effective irradiance in $\mu W/cm^2$ [$\mu J/(s \cdot cm^2)$] or W/m^2 [$J/(s \cdot m^2)$] normalized to a monochromatic source at 270 nm

E_λ = spectral irradiance from measurements in $\mu W/(cm^2 \cdot nm)$ or $W/(m^2 \cdot nm)$

S_λ = relative spectral effectiveness (unitless)

Δ_λ = bandwidth in nanometers of the calculation or measurement intervals.

Permissible exposure time in seconds for exposure to actinic UVR incident upon the unprotected skin or eye may be computed by dividing 30 J/m^2 by the value of E_{eff} in W/m^2. The maximal exposure duration may also be determined using Table 2, which provides representative exposure durations corresponding to effective irradiances in W/m^2 or $\mu W/cm^2$.

SPECIAL CONSIDERATIONS

These EL values are intended to apply to UVR exposure of the working population, but with some precaution also apply to the general population. However, it should be recognized that some rare, highly photosensitive individuals exist who may react adversely to exposure at these levels. These individuals are normally aware of their heightened sensitivity. Likewise, if individuals are concomitantly exposed to photosensitizing agents (see, for example, reference Fi74), a photosensitizing reaction can take place. It should be emphasized that many individuals who are exposed to photosensitizing agents (ingested or externally applied chemicals, e.g., in cosmetics, foods, drugs, industrial chemicals, etc.) probably will not be aware of their heightened sensitivity. Lightly pigmented individuals conditioned by previous UVR exposure (leading to tanning and hyperplasia) and heavily pigmented individuals can tolerate skin exposure in excess of the EL without erythemal effects. However, repeated tanning may increase the risk for those persons later developing signs of accelerated skin aging and even skin cancer. Such risks should be understood prior to the use of UVR for medical phototherapy or cosmetic exposures.

TABLE 1. *UV radiation exposure limits and spectral weighting function. IRPA/INIRC 1988 Revision (IR89)*

Wavelength[a] (nm)	EL ($J\ m^{-2}$)	EL ($mJ\ cm^{-2}$)	Relative Spectral Effectiveness S_λ	Wavelength[a] (nm)	EL ($J\ m^{-2}$)	EL ($mJ\ cm^{-2}$)	Relative Spectral Effectiveness S_λ
180	2,500	250	0.012	310	2,000	200	0.015
190	1,600	160	0.019	313[b]	5,000	500	0.006
200	1,000	100	0.030	315	1.0×10^4	1.0×10^3	0.003
205	590	59	0.051	316	1.3×10^4	1.3×10^3	0.0024
210	400	40	0.075	317	1.5×10^4	1.5×10^3	0.0020
215	320	32	0.095	318	1.9×10^4	1.9×10^3	0.0016
220	250	25	0.120	319	2.5×10^4	2.5×10^3	0.0012
225	200	20	0.150	320	2.9×10^4	2.9×10^3	0.0010
230	160	16	0.190	322	4.5×10^4	4.5×10^3	0.00067
235	130	13	0.240	323	5.6×10^4	5.6×10^3	0.00054
240	100	20	0.300	325	6.0×10^4	6.0×10^3	0.00050
245	83	8.3	0.360	328	6.8×10^4	6.8×10^3	0.00044
250	70	7.0	0.430	330	7.3×10^4	7.3×10^3	0.00041
254[b]	60	6.0	0.500	333	8.1×10^4	8.1×10^3	0.00037
255	58	5.8	0.520	335	8.8×10^4	8.8×10^3	0.00034
260	46	4.6	0.650	340	1.1×10^5	1.1×10^4	0.00028
265	37	3.7	0.810	345	1.3×10^5	1.3×10^4	0.00024
270	30	3.0	1.000	350	1.5×10^5	1.5×10^4	0.00020
275	31	3.1	0.960	355	1.9×10^5	1.9×10^4	0.00016
280[b]	34	3.4	0.880	360	2.3×10^5	2.3×10^4	0.00013
285	39	3.9	0.770	365[b]	2.7×10^5	2.7×10^4	0.00011
290	47	4.7	0.640	370	3.2×10^5	3.2×10^4	0.000093
295	56	5.6	0.540	375	3.9×10^5	3.9×10^4	0.000077
297[b]	65	6.5	0.460	380	4.7×10^5	4.7×10^4	0.000064
300	100	10	0.300	385	5.7×10^5	5.7×10^4	0.000053
303[b]	250	25	0.190	390	6.8×10^5	6.8×10^4	0.000044
305	500	50	0.060	395	8.3×10^5	8.3×10^4	0.000036
308	1,200	120	0.026	400	1.0×10^6	1.0×10^5	0.000030

[a] Wavelengths chosen are representative; other values should be interpolated at intermediate wavelengths.
[b] Emission lines of a mercury discharge spectrum.

TABLE 2. *Limiting UV exposure durations based on exposure limits*

Duration of exposure per day	Effective irradiance	
	E_{eff} (W/m^2)	E_{eff} (μW/cm^2)
8 hours	0.001	0.1
4 hours	0.002	0.2
2 hours	0.004	0.4
1 hour	0.008	0.8
30 minutes	0.017	1.7
15 minutes	0.033	3.3
10 minutes	0.05	5
5 minutes	0.1	10
1 minute	0.5	50
30 seconds	1.0	100
10 seconds	3.0	300
1 second	30	3,000
0.5 second	60	6,000
0.1 second	300	30,000

PROTECTIVE MEASURES

Protective measures will differ depending upon whether the UVR exposure occurs indoors or outdoors. The use of hats, eye protectors, facial shields, clothing, and sun-shading structures are practical protective measures. As with any indoor, industrial hazard, engineering control measures are preferable to protective clothing, goggles, and procedural safety measures. Glass envelopes for arc lamps will filter out most UV-B and UV-C. Where lengthy exposure to high-power glass-envelope lamps, and quartz halogen lamps will occur at close proximity, additional glass filtration may be necessary. Light-tight cabinets and enclosures and UVR absorbing glass and plastic shielding are the key engineering control measures used to prevent human exposure to hazardous UVR produced in many industrial applications such as the fade testing of materials, solar simulation, photoresist applications, and photocuring. For arc welding, cabinets are not practical. Shields, curtains, baffles, and a suitable separation distance are used to protect individuals against the UVR emitted by open-arc processes such as arc welding, arc-cutting, and plasma spraying. Progress is being made in the development of dynamic-filter welding helmets, see-through curtains, and other new safety equipment. There is a need for operational rules to protect potentially exposed individuals. Operators should be trained to follow these general rules properly. Ventilation may be required exhausting ozone and other airborne contaminants produced by UV-C radiation.

MEASUREMENT

Although UVR radiometers exist, attempts to produce relatively inexpensive field safety survey meters that respond directly to UV-B and UV-C radiation (following the S_λ function) have not been fully successful. However, relatively expensive instruments exist that respond to UV-B and UV-C radiation according to the relative spectral effectiveness, S_λ. Spectroradiometric measurements of the source that can then be used with the S_λ weighting function to calculate E_{eff} are often necessary for measurements more accurate than with simple, direct-reading safety meters. Whichever measurement technique is applied, the geometry of measurement is important. All the preceding ELs for UVR apply to sources that are measured with an instrument having a cosine-response detector oriented perpendicular to the most directly exposed surfaces of the body when assessing skin exposure and along (or parallel to) the line(s) of sight of each exposed individual when assessing ocular exposure. The use of UV film badges makes it possible to integrate UV exposure on specific body sites that move with respect to the UVR source; however, the spectral response of such film badges still does not accurately follow S_λ.

CONCLUDING REMARKS

The increasing use of UVR in medicine, in the industrial work environment, for cosmetic use, and partly in consumer products necessitates that greater attention be paid to the potential hazards of this type of electromagnetic radiation. The present understanding of chronic effects and injury mechanisms of UVR is limited, and this problem awaits further research. The above guidelines will be subject to periodic review and amendment as appropriate.

REFERENCES

Ba82 Bauer H., Caldwell M. M., Tevini M. and Worrest B. (Eds.), 1982, *Biological effects of UV-B radiation* (Proc. of a workshop held in Munich-Neuherberg, 25–27 May 1982), Gesellschaft für Strahlen- und Umweltforschung, Josephspitalstr. 15, D-8000 Munich 2, Federal Republic of Germany, BPT-Report 5/82.

Be68 Berger D., Urbach F. and Davies R. E., 1968, "The action spectrum of erythema induced by ultraviolet radiation" (Preliminary Report XIII), pp. 1112–1117, in: *Proc. of the Congressus Internationalis Dermatologiae—München* 1967 (Edited by W. Jadassohn and C. G. Schirren) (New York: Springer-Verlag).

Be70 Bernstein H. N., Curtis J., Earl F. L. and Kuwabara T., 1970, "Phototoxic corneal and lens opacities," *Arch. Ophthalmol.* **83,** 336–348.

CIE70 Commission Internationale de l'Eclairage (International Commission on Illumination), 1970, *International Lighting Vocabulary,* 3rd Edn, Publication CIE No. 17 (E-1.1) (Paris: CIE).

Ch78 Chavaudra J. and Latarjet R., 1978, "Le rayonnement ultraviolet solaire et l'ozone atmosphérique," in: *Effets biologiques des rayonnements non-ionisants—utilisation et risques associés* (Proc. 9th Int. Congr. Société française de Radioprotection) (B.P. no. 72, F-92260, Fontenay-aux-Roses, France: SFRP) (in French).

Co31 Coblentz W. W., Stair R. and Hogue J. M., 1931, "The spectral erythemic reaction of the human skin to ultraviolet radiation," *Proc. U.S. Nat. Acad. Sci.* **17,** 401–403.

Cu74 Cutchis P., 1974, "Stratospheric ozone depletion and solar ultraviolet radiation on earth," *Science* **184,** 13–19.

Cu77 Cunningham-Dunlop S., Kleinstein B. H. and Urbach F., 1977, *A Current Literature Report on the Carcinogenic Properties of Ionizing and Non-ionizing Radiation:* **I.** *Optical Radiation,* National Institute for Occupational Safety and Health, Cincinnati, OH, Contract 210-76-0145, DHEW (NIOSH) Publication No. 78-122.

De78 Despres S., 1978, "Effets biologiques des infrarouges et des ultraviolets," *Radioprotection* **13(1),** 11–21 (in French).

Ev65 Everett M. A., Olsen R. L. and Sayer R. M., 1965, "Ultraviolet erythema," *Arch. Dermatol.* **92,** 713–729.

Fi35 Fischer F. P., Vermeulen D. and Eymers J. G., 1935, "Über die zur Schädigung des Auges nötige Minimalquantität von ultraviolettem und infrarotem Licht," *Arch Augenheilk.* **109,** 462–467 (in German).

Fi74 Fitzpatrick T. B., Pathak M. A., Harber L. C., Seiji M. and Kutika A. (Eds.), 1974, *Sunlight and Man* (Tokyo: University of Tokyo Press).

Fr66 Freeman R. G., Owens D. W., Knox J. M. and Hudson H. T., 1966, "Relative energy requirements for an erythemal response of skin to monochromatic wavelengths of ultraviolet present in the solar spectrum," *J. Invest. Dermatol.* **47,** 586–592.

Ge78 Gezondheidsraad (Health Council of the Netherlands), 1978, *Recommendations Concerning Acceptable Levels of Electromagnetic Radiation in the Wavelength Range from* 100 nm *to* 1 mm *(Micrometre Radiation),* Ministry of Health and Environmental Protection, Postbox 439-2260 AK Leidschendam, The Netherlands, Report 65E (March).

Go76 Gordon D. and Silverstone H., 1976, "Worldwide epidemiology of pre-malignant and malignant cutaneous lesions," pp. 405–434, in: *Cancer of the Skin* (Philadelphia: W. B. Saunders Co.).

Ha28 Hausser K. W., 1928, "Influence of wavelength in radiation biology," *Strahlentherapie* **28,** 25–44 (in German).

Ha69a Hamerski W. and Zajaczkowska A., 1969, "Electrophoretic investigations of proteins of the corneal epithelium in experimental photophthalmia," *Pol. Med. J.* **8,** 1464–1468.

Ha69b Hamerski W., 1969, "Studies on the histochemical changes in experimental corneal lesions induced with ultraviolet radiation and on prevention of photophthalmia," *Klin. Oczna* **39,** 537–542 (in Russian). English translation in: *Pol. Med. J.* **8,** 1469–1476.

Ha82 Ham W. T. Jr., Mueller H. A., Ruffolo J. J. Jr., Guerry D. III and Guerry R. K., 1982, "Action spectrum for retinal injury from near-ultraviolet radiation in the aphakic monkey," *Amer. J. Ophthalmol.* **93,** 299–306.

Hi77 Hiller R., Giacometti L. and Yuen K., 1977, "Sunlight and cataract: an epidemiological investigation," *Am. J. Epidemiol.* **105,** 450–459.

IR89 International Radiation Protection Association/International Non-Ionizing Radiation Committee, 1989, "Proposed change to the IRPA 1985 guidelines on limits of exposure to ultraviolet radiation", *Health Phys.* **56(6),** 971–972.

Ku77 Kurzel R. B., Wolbarsht M. L. and Yamanashi B. S., 1977, "Ultraviolet radiation effects on the human eye," pp. 133–168, in: *Photochemical and Photobiological Reviews,* Vol. 2 (Edited by K. C. Smith) (New York: Plenum Press).

Lu30 Luckiesh M. L., Holladay L. and Taylor A. H., 1930, "Reaction of untanned human skin to ultraviolet radiation," *J. Opt. Soc. Amer.* **20,** 423–432.

Ma72 Marshall J., Mellerio J. and Palmer D., 1972, "Damage to pigeon retinae by moderate illumination from fluorescent lamps," *Exp. Eye Res.* **14,** 164–169.

Ma79 Mayer M. A. and Salsi M. S., 1979, *Rayonnement ultraviolet, visible et infrarouge: mesure-évaluation des risques,* Institut national de Recherche et de Sécurité, 30 rue Olivier-Noyer, 75014, Paris, France, Cahiers de Notes Documentaires, Note No. 1199-96-79, No. 96 (in French).

Mc87 McKinlay A. F. and Diffey B. L., 1989, "A reference action spectrum for ultraviolet induced erythema in human skin," in: *Human exposure to ultraviolet radiation: risks and regulations* (Edited by W. F. Passchier, B. F. M. Bosnjakovic). Proceedings of a seminar held in Amsterdam, March 23–25, 1987, pp. 83–87 (Amsterdam: Excerpta Medica, Elsevier Science Publishers B. V.).

Pa78 Parrish J. A., Anderson R. R., Urbach F. and Pitts D., 1978, *UV-A. Biological Effects of Ultraviolet Radiation with Emphasis on Human Responses to Longwave Ultraviolet* (New York: Plenum Press).

Pa82 Parrish J. A., Jaenicke K. F. and Anderson R. R., 1982, "Erythema and melanogenesis action spectra of normal human skin," *Photochem. Photobiol.* **36(2),** 187–191.

Pau82 Paul B. and Parrish J., 1982, "The interaction of UV-A and UV-B in the production of threshold erythema," *J. Invest. Derm.* **78,** 371–374.

Pi71 Pitts D. G. and Tredici T. J., 1971, "The effects of ultraviolet on the eye," *Amer. Ind. Hyg. Assoc. J.,* **32(4),** 235–246.

Pi74 Pitts D. G., 1974, "The human ultraviolet action spectrum," *Amer. J. Optom. and Physiol. Optics* **51(12),** 946–960.

Pi77 Pitts D. G., Cullen A. P. and Hacker P. D., 1977, *Ocular Ultraviolet Effects from* 295 nm *to* 400 nm *in the Rabbit Eye,* National Institute for Occupational Safety and Health, Cincinnati, OH, Contract CDC-99-74-12, DHEW (NIOSH) Publication No. 77-175.

Pir71 Pirie A., 1971, "Formation of N-formylkynurenine in proteins from lens and other sources by exposure to sunlight," *Biochem. J.* **125,** 203–208.

Sc64 Schmidt K., 1964, "On the skin erythema effect of UV flashes," *Strahlentherapie* **124,** 127–136.

Sh77 Sherashov S. G., 1977, "Spectral sensitivity of the cornea to ultraviolet radiation," *Biofizika* **15,** 543–544 (in Russian).

Sl72 Sliney D. H., 1972, "The merits of an envelope action spectrum for ultraviolet exposure criteria," *Am. Ind. Hyg. Assoc. J.* **33,** 644–653.

Sl80 Sliney D. H. and Wolbarsht M. L., 1980, *Safety with Lasers and Other Optical Sources. A Comprehensive Handbook* (New York: Plenum Press).

So64 Sollner F., 1964, "Über die Lichtabsorption eiweissfreier Extrakte von frischen und konservierten Hornhauten," *Albrecht von Graefes Arch. Ophthalmol.* **167,** 527–536 (in German).

Sw71 Swanbeck G. and Hillström L., 1971, "Analysis of etiological factors of squamous cell skin cancer of different locations," *Acta Dermatovener.* **51,** 151–156.

Ta69 Tapaszto I. and Vass Z., 1969, "Alterations in mucopolysaccharide compounds of tear and that of corneal epithelium, caused by ultraviolet radiation," *Ophthalmologica* (Additamentum) **158,** 343–347.

Tr25 Trumpy E., 1925, "Experimentelle Untersuchungen über die Wirkung hochintensiven ultravioletts und violetts zwischen 354 und 435.9 millimicrons Wellenlänge auf das Auge unter besonderer Berücksichtigung der Linse," *Albrecht von Graefes Arch. Ophthalmol.* **115,** 495–514 (in German).

UN79 United Nations Environment Programme/World Health Organization/International Radiation Protection Association, 1979, *Environmental Health Criteria* 14, *Ultraviolet Radiation* (Geneva: WHO).

Ur74 Urbach F., Epstein J. H. and Forbes P. D., 1974, "UV carcinogenesis," pp. 259–283, in: *Sunlight and Man* (Edited by T. B. Fitzpatrick, M. A. Pathak, L. C. Harber, M. Seiji and A. Kutika) (Tokyo: University of Tokyo Press).

Ur86 Urbach F., and Gange R. W., 1986, *The biological effects of UVA radiation.* (New York: Praeger Publications).

Va25 van der Hoeve J., 1925, "Strahlen und Auge," *Albrecht von Graefes Arch. Ophthalmol.* **116,** 245–248 (in German).

Va69 Van der Leun J. C. and Stoop T., 1969, in: *The Biological Effects of UV Radiation* (Edited by F. Urbach), pp. 251–254 (Oxford: Pergamon Press).

Va84 Van der Leun J. C., 1984, "UV carcinogenesis," *Photochem. Photobiol.* **39(6),** 861–868.

Wi72 Willis I., Kligman A. and Epstein J., 1972, "Effects of long ultraviolet rays on human skin: photoprotective or photoaugmentative," *J. Invest. Dermatol.* **59,** 416–420.

Zu77 Zuclich J. A. and Kurtin W. E., 1977, "Oxygen dependence of near UV-induced corneal damage," *Photochem. Photobiol.* **25,** 133–135.

Zu80 Zuclich J. A., 1980, "Cumulative effects of near-UV induced corneal damage," *Health Phys.* **38,** 833–838.

APPENDIX: RATIONALE FOR THE LIMITS OF EXPOSURE TO UVR

Background

A comprehensive review of UVR effects has been published by UNEP/WHO/IRPA (UN79) and the interested reader is referred to that document. The following discussion is a brief review of those physical and biological factors used to derive the UVR guidelines.

General biological effects

Life has evolved under the daily exposure to solar radiation. Although UVR is only about 5% of the sunlight that reaches the earth's surface, it plays a significant biological role since the energies of individual photons are the greatest for any of the photons in the solar spectrum. These shorter wavelength, higher energy photons have sufficient energy to initiate biological effects that may be injurious. The critical organs for UVR exposure are the eye and the skin since they may be readily exposed. In some dental applications, the interior of the mouth may also be exposed.

The thresholds for the observed bioeffects vary significantly with wavelength. Consequently, various "action spectra" have been developed to establish dose-response relationships. In photobiology, the term "action spectrum" refers to the relative spectral effectiveness of different wavelengths in eliciting a biological effect.

Erythema

Erythema (e.g., the reddening of the skin in sunburn) is the most commonly observed effect on skin after exposure to UVR. This effect was first quantitatively documented as a wavelength dependent effect in the late 1920s by Hausser and Vahle in Germany (Ha28). These and other quantitative studies since that time have confirmed that the erythemal threshold varies with anatomical site, wavelength, and time between exposure and assessment (Ba82; Be68; Co31; Ev65; Fr66; Lu30; Pa82; Va69; Wi72). In addition, the variation in published threshold values is due to differences in the clinical definition and estimate of minimal erythema and radiometric measurement techniques. Erythema is a photochemical response of the skin normally resulting from overexposure to wavelengths in the UV-C and UV-B regions (180–315 nm). Exposure to UV-A alone can produce erythema, but only at very high radiant exposures (>10 J/cm^2) (Pa78). The UV-A added prior to UV-B exposure may slightly intensify the erythemal response (Wi72). This synergistic effect of two spectral bands is known as photoaugmentation. The opposite effect where one previous exposure desensitizes the skin also occurs, and may be more pronounced for simultaneous exposure of UV-A and UV-B (Pau82; Va69). Hausser and Vahle first showed (as reported by Hausser) that erythema induced by the longer UV-B wavelengths (295–315 nm) is more severe and persists for a longer period than that for shorter wavelengths (Ha28). The increased severity and time course of the erythema may result from the deeper penetration of these wavelengths into the epidermis. In general it is accepted that UVR releases a number of diffusing mediators, which in turn carries the inflammatory effect into deeper skin layers. The assessment of a threshold for the maximum sensitivity of the skin to erythema varies from 250 to 297 nm depending upon the criteria of assessment and the period following the exposure. Action spectra for different grades of erythema are quite different. For the most severe grade of erythema this maximum sensitivity occurs between 290 and 300 nm. The minimal erythema dose (MED) reported in more recent studies for untanned, lightly pigmented skin range from 6 to 30 mJ/cm^2 (Ev65; Fr66; Pa82). These MED data suggest that for this type of skin, the EL values are approximately 1.3 to 6.5 times less than the MED values. Skin pigmentation and "conditioning" (thickening of the stratum corneum and tanning) may result in an increase of the MED by at least one order of magnitude. Figure 1 shows the variation of the skin erythema action spectrum.

Delayed effects on the skin

Chronic exposure to sunlight, especially the UV-B component, accelerates the skin aging process and increases the risk of developing skin cancer (Va84). The solar spectrum is greatly attenuated by the earth's ozone layer, limiting terrestrial UV to wavelengths greater than approximately 290 nm. The UV-B irradiance at ground level is a strong function of the sun's elevation angle in the sky. This is due to the change of UV attenuation with atmospheric path length (time of day and season). Several epidemiologic studies have shown that the incidence of skin cancer is strongly correlated with latitude, altitude, and cloud cover (Ch78; Cu74; Cu77; Go76; Sw71; Ur74). Exact quantitative dose-response relationships have not yet been established although fair-skinned individuals, especially of Celtic origin, are much more prone to develop skin cancer. Skin cancer is typically a disease of outdoor workers such as farmers and seamen (Ur74). Only a few quantitative studies have examined work populations chronically exposed to artificial sources of UV-B to determine whether there is an increased skin cancer risk in the occupational environment. Squamous cell carcinoma is the most common type. This is localized at exposed sites (e.g., hands and back of the neck). No studies of the incidences of melanoma have been reported for outdoor workers.

Photokeratoconjunctivitis

Actinic UVR (UV-B and UV-C) is strongly absorbed by the cornea and conjunctiva. Overexposure of these tissues causes photokeratoconjunctivitis, commonly referred to as welder's flash, arc-eye, etc. Pitts has characterized the course of ordinary clinical photokeratitis (Pi71; Pi74; Pi77). The latent period varies inversely with the severity of exposure ranging from 1/2 to 24 hours but usually occurs within 6–12 hours. Conjunctivitis tends to develop more slowly and may be accompanied by erythema of the facial skin surrounding the eyelids. The individual has the sensation of a foreign body or sand in the eyes and may experience photophobia, lacrimation, and blepharospasm to varying degrees. The acute symptoms last from 6 to 24 hours and discomfort usually disappears within 48 hours.

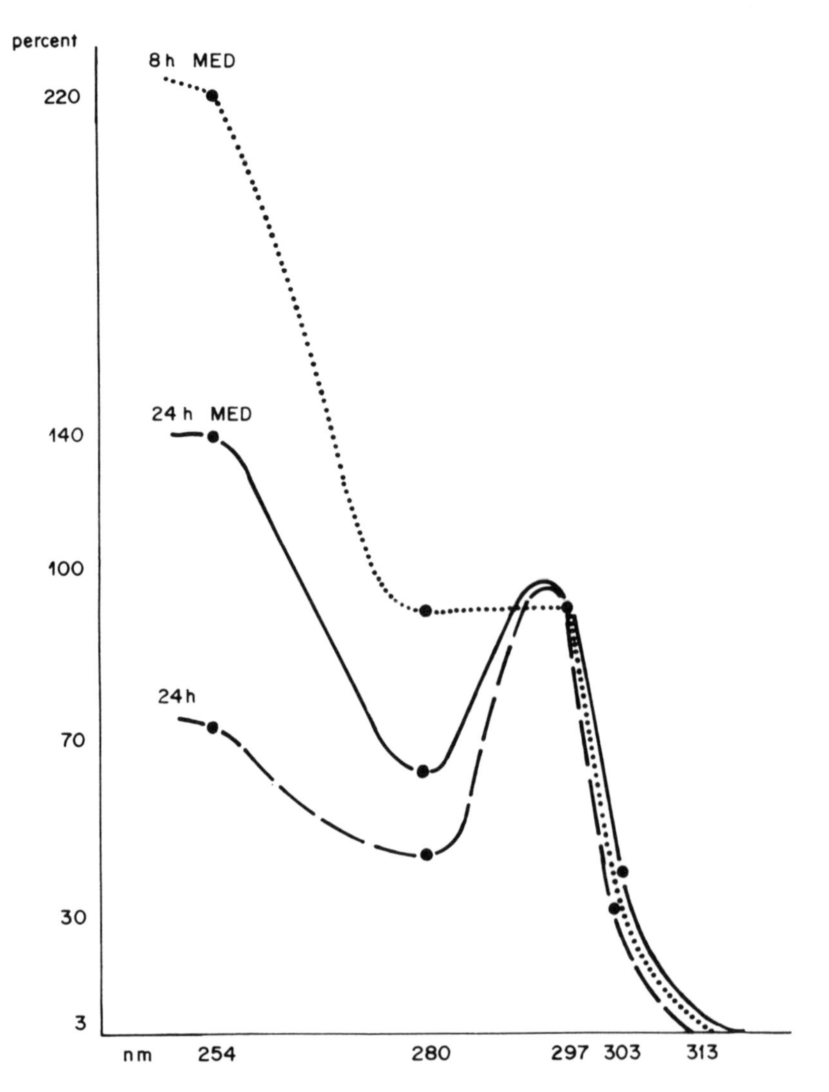

Figure 1. Action spectrum of human skin. Averages of relative values for abdominal skin of five subjects. Note great similarity for wavelengths from 297–313 nm, and marked differences at shorter wavelengths for 8 hours (after irradiation) MED, 24 hours (after irradiation) MED, and a curve constructed by using values for moderate erythema.

Although exposure rarely results in permanent ocular injury, the individual is visually incapacitated during this 48-hour period. Threshold data for photokeratitis in humans have been established by Pitts and Tredici for 10 nm wavebands from 220 to 310 nm (Pi71). The guideline ELs between 200 nm and 305 nm are about 1.3 to 4.6 times less than the threshold for minimal change. The maximum sensitivity of the human eye was found to occur at 270 nm. The wavelength response (action spectrum) between 220 and 310 nm does not vary as greatly as in the case of erythema with the thresholds varying from 4 to 14 mJ/cm^2. Corneal injury from UV-A wavelengths requires levels exceeding 10 J/cm^2 (Ha69b; Pi77; Sh77; Ta69; Zu77; Zu80).

Cataract

Wavelengths above 295 nm can be transmitted through the cornea and are absorbed by the lens. Pitts et al (Pi77) have shown that both transient and permanent opacities of the lens (cataracts) can be produced in rabbits and monkeys by exposure to UVR having wavelengths in the 295–320 nm band. Thresholds for transient opacities ranged dramatically with wavelength, from 0.15 to 12.6 J/cm^2. Thresholds for permanent opacities were typically twice those for transient opacities (Pi77). Whether chronic exposure at lower levels will produce lenticular opacities has not been determined (Fi35; Ha69b; Hi77; Ku77; Ma72; Pi71; Pir71; So64; Tr25; Va25).

Retinal effects

The cornea and crystalline lens normally shield the retina sufficiently from acute effects from UVR exposure. Normally, less than 1% of UV-A reaches the retina, shorter UV-B wavelengths being totally attenuated (Pi77; Sl80). Upon removal of the crystalline lens, Ham and colleagues (Ha82) demonstrated acute retinal injury at levels of the order of 5 J/cm^2 at the retina.

Envelope action spectrum

a. Clearly, the development of UVR exposure limits for workers and the general population must consider two risks. These are the risks of acute and chronic injury to both the eye and skin. The literature indicates that thresholds for injury vary significantly with wavelength for each effect. In the UV-B and UV-C regions, an action spectrum curve can be drawn that envelopes the threshold data for exposure doses (radiant exposures) in the range of reciprocity (Sc64; Zu80) for acute effects obtained from recent studies of minimal erythema and keratoconjunctivitis. Reciprocity means that irradiance E and exposure duration t have a reciprocal relation, and a constant product of E and t (i.e., exposure) results in a given effect. This EL curve does not differ significantly from the collective threshold data considering measurement errors and variations in individual response (Sl72; Sl80). Although the safety factor is minimal for minimally detectable keratitis, it is believed to range from 1.5 to 2.0 for acute keratitis. The curve is also well below the acute UV-B cataractogenic thresholds (Pi77; Sl80). Repeated exposure of the eye to potentially hazardous levels of UV is not believed to increase the protective capability of the cornea as does skin tanning and thickening of the stratum corneum. Thus, this EL is more readily applicable to the eye and must be considered a limiting value for that organ (Sl72). Any accumulation of UV-B and UV-C exposures causing photokeratitis is limited to about 48 hours since the outer corneal epithelial layers are replaced in about 48 hours by the normal repair process of this tissue. Some slight additivity of UV-A exposures exists beyond 48 hours because of the deeper penetration of UV-A rays (Zu80). The additivity factors were considered in deriving the magnitude of the safety factor built into the EL. On the basis of acute effects, the safety factor for UV-A ELs is large, varying from about 7 at 320 nm to more than 100 at 390 nm.

b. Because of the wide variations in threshold values and exposure history (conditioning) among individuals, these guidelines should only be used as a starting point for evaluating skin hazards (De78; Ge78; Ma79; Sl80). The envelope guideline has some margin of safety to protect all but the most sensitive individuals. An exact value for this margin cannot be given, but for lightly pigmented persons, it varies from about 3 to 20 depending on the spectral composition of the radiation. Since there may be more than one erythemal mechanism and, therefore, more than one erythemal action spectrum, the effect of radiations of two widely differing wavelengths in the 180 nm–315 nm range may not be simply additive. The EL should be used with caution in evaluating sources such as the sun and fluorescent lamps, having a rapidly increasing spectral irradiance in the 300–315 nm range. Large errors can arise because of the difficulty in making accurate spectral measurements of such sources in this region.

c. The EL may not provide adequate protection for photosensitive individuals or for normal individuals exposed concomitantly to chemical, pharmaceutical, or phyto-photosensitizers, and special precautions must be taken for such cases (Be70, Pa78).

d. The EL should reduce the risk of occurrence of chronic skin effects by preventing acute effects and limiting lifelong UV exposure. An action spectrum for UV skin carcinogenesis is not known for man, although the erythemal action spectrum has been used for global estimates of UV exposure. The Dutch Health Council (Ge78) proposed envelope limits similar to these ELs for acute daily exposure up to 10 years duration. For longer periods, to protect against chronic effects, another set of values that are about 15% of the acute set should be used. These should have the same action spectrum in the UV-B and UV-C. In many cases, occupational exposure to UV-B adds to an individual's nonoccupational exposure to solar UV-B.

e. In addition to the UV hazard, very intense UV-C sources may also produce hazardous concentrations of ozone and nitrogen oxides from the air and of phosgene gas in the presence of degreasers.

UV-A radiation effects

Recent studies of skin and ocular injury action spectra in the near-UV spectral region (315–400 nm) provided sufficient data (Ba82; Mc87; Pa82; Ur86) to update the exposure limits for the unprotected eye in the UV-A region and to expand the values of the relative spectral effectiveness, S_λ, up to 400 nm for determining the limits of exposure of the skin to UV-A radiation. However, few industrial sources emit sufficient intensity in this spectral region to cause adverse biologic effects. Except for a few "black-light" sources, measurements of most broadband lamps and of welding arcs will not be affected by the new EL in terms of effective irradiance. Skin damage is principally thermal in nature requiring very high irradiances except in photosensitive individuals (Pa82). Photokeratitis and lenticular opacities have been produced in experimental animals with acute exposure at high radiant exposures (Pi77; Zu77; Zu80). There are no indications that the low levels of UV-A found in most indoor work environments present a hazard although it has been hypothesized as one causative agent for cataracts in the past (Sl80). Thus, the EL for UV-A should be below most conceivable thermal or photochemical injury mechanisms.

In recent years there has been a rapidly growing population of aphakic individuals who have had one or both crystalline lenses removed following cataract. Many of these aphakics have received artificial intraocular lenses of glass or plastic. (Such individuals are frequently referred to as pseudophakics). Aside from a few with implants designed to absorb UV-A, these persons would not be adequately protected against retinal injury from UV-A exposure at the EL (Ha82). Such persons should be fitted with UV-A protective lenses if working with sources of UV-A radiation.

CHAPTER 4

GUIDELINES ON LIMITS OF EXPOSURE TO LASER RADIATION OF WAVELENGTHS BETWEEN 180 nm AND 1 mm

A DOCUMENT entitled *Environmental Health Criteria 23, Lasers and Optical Radiation* (UN82) was published in 1982 under the joint sponsorship of the United Nations Environment Programme, the World Health Organization, and the International Radiation Protection Association (IRPA). The document contains a review of the biological effects reported from exposure to optical radiation from lasers and other optical sources and serves as the scientific rationale for the development of these guidelines.

The purpose of these guidelines is to deal with the basic principles of protection against optical radiation emitted by lasers, so that they may serve as guidance to the various international and national bodies or individual experts who are responsible for the development of regulations, recommendations, or codes of practice to protect the workers and the general public from the potentially adverse effects of laser radiation.

The committee recognized that when standards on exposure limits are established, various value judgments are made. The validity of scientific reports has to be considered, and extrapolations from animal experiments to effects on humans have to be made. Cost vs. benefit analyses are necessary, including economic impact of controls. The limits in these guidelines were based on the scientific data and no consideration was given to economic impact or other nonscientific priorities. However, from presently available knowledge, the limits should provide an adequate level of protection against known laser effects under all normal exposure conditions.

The IRPA Associate Societies as well as a number of competent institutions and individual experts were consulted in the preparation of these guidelines and their cooperation is gratefully acknowledged.

The guidelines on limits of exposure to laser radiation were first approved by the IRPA Executive Council in May 1984 and published in *Health Physics* in August 1985. They were subsequently updated as the data base of laser biological effects upon the human eye and skin had improved (*Health Physics,* May 1988). During the preparation of the initial guidelines, the composition of the International Non-Ionizing Radiation Committee (INIRC) of the IRPA was as follows:

H. P. Jammet, Chairman (France)
B. F. M. Bosnjakovic (Netherlands)
P. Czerski (Poland)
M. Faber (Denmark)
D. Harder (Germany)
J. Marshall (Great Britain)
M. H. Repacholi (Australia)
D. H. Sliney (U.S.A.)
J. C. Villforth (U.S.A.)
A. S. Duchêne, Scientific Secretary (France)

In addition, the following members participated in the guidelines later updating: J. Bernhardt (Germany), M. Grandolfo (Italy), B. Knave (Sweden), J. A. J. Stolwijk (U.S.A.).

INTRODUCTION

Lasers are used in a wide variety of industrial, consumer, scientific, and medical applications, including alignment, welding, cutting, drilling, heat treatment, distance measurement, entertainment, advertisement, and surgery. In most industrial laser applications the laser radiation is totally enclosed, or alternatively partial enclosures effectively preclude direct human exposure. However, in some applications, human exposure to potentially hazardous laser radiation is possible; for example, in research laboratories, laser displays, and alignment procedures. Other industrial material processing applications use high intensities that can produce sufficient scattered radiation to have adverse health implications. Additionally, many people are exposed to levels of laser radiation that are not known to produce biological damage such as that emitted by laser uniform-product-code (UPC) scanners in supermarkets or that which is allowed in audience areas at laser light shows.

Increasing numbers and varieties of consumer and office devices that incorporate lasers include video disc players, supermarket scanners, facsimile and printing equipment, and guidance devices for blind people. In general, these applications employ low-intensity laser radiation in the wavelength range between 630 nm and 910 nm (red light and near-infrared radiation).

Adverse health effects of exposure to laser radiation are of particular concern in the visible and near infrared (400 nm–1400 nm) where retinal injury can occur, although adverse biological effects are theoretically possible across the entire optical spectrum from 180 nm in the ultraviolet (UV) to 1 mm (10^6 nm) in the far infrared (IR). Within this region exposure limits vary enormously because of variations in biological effects and the different critical structures of the eye potentially at risk. Exposure limits (ELs) have been developed for this entire wavelength region for exposure durations between 1 ns and 8 hours. The biological effects induced by optical radiation are essentially the same for coherent and incoherent sources for any given wavelength, exposure area, and duration, however there is a necessity to treat lasers as a special case because few conventional optical sources can approach the radiant intensities and irradiances achieved by lasers. Furthermore, because many of the early data points for biological effects were developed using conventional optical sources with emission of radiation over a broadband of wavelengths, these data are not directly applicable to the highly monochromatic emissions of lasers. The degree of uncertainty in relating biological thresholds derived from broadband and monochromatic sources has frequently led to added safety factors in the ELs for lasers, and this is particularly the case in the UV range.

SCOPE AND PURPOSE

The ELs listed herein apply to the wavelength range from 180 nm to 1 mm and are based on an international consensus on the health effects and hazards, as expressed in the health criteria document quoted above. These guidelines apply to all human exposure, both acute and chronic, to optical radiation emitted by lasers, except for deliberate exposures to patients undergoing medical treatment in which the exposure is part of the therapeutic procedure. In general, diagnostic procedures should not exceed these ELs. Any exposure above the EL would necessitate a risk vs. benefit analysis as such exposure could result in permanent damage.

BASIC CONCEPTS

A laser (an acronym for Light Amplification by the Stimulated Emission of Radiation) is a device that emits electromagnetic radiation having wavelengths in the optical region (100 nm–1 mm). The radiation emitted by a laser is coherent. A laser beam is typically monochromatic and highly collimated as compared with the output of conventional optical sources. Because of these special properties, radiometric measurement techniques and potential exposure conditions may vary significantly from those of conventional optical sources. For these reasons, exposure limits for lasers are specialized and incorporate assumptions of exposure that may not apply to conventional optical sources. Hence, their applicability to the assessment of health risks from conventional light sources is limited and should not be undertaken without a thorough knowledge of the assumptions underlying the ELs.

Electromagnetic radiation in the wavelength range between approximately 100 nm and 1 mm is now widely and increasingly termed "optical radiation." In the wavelength range between 100 nm and 400 nm radiation is termed ultraviolet radiation (UVR). From 400 to 760 nm it is termed light (or visible radiation). Radiation having wavelengths from 760 nm to 1 mm is termed infrared radiation (IR). The UV and IR regions may be further subdivided: UV-C from 100 to 280 nm, UV-B from 280 to 315 nm, UV-A from 315 to 400 nm, IR-A from 760 to 1,400 nm, IR-B from 1,400 to 3,000 nm and IR-C from 3,000 nm to 1 mm (1,000,000 nm). These spectral band limits defined by the International Commission on Illumination (CIE) are useful in discussing the biological effects of optical radiation. However, the predominant biological effects have much less sharply defined spectral limits.

Quantities and units

Optical radiation protection ELs are expressed using the following quantities: Irradiance (E) expressed in W/m^2, mW/cm^2, or $\mu W/cm^2$, and radiant exposure (H) in J/m^2, mJ/cm^2, or $\mu J/cm^2$ are used in describing the concepts of dose rate and dose from direct laser radiation. Radiance (L) expressed in $W/(m^2 \cdot sr)$ or $W/(cm^2 \cdot sr)$ and time-integrated radiance (Lp) expressed in $J/(m^2 \cdot sr)$ or $J/(cm^2 \cdot sr)$ are used to describe the "brightness" of an extended source that may be imaged on the retina.

Limiting apertures and fields-of-view for measurement

Radiant exposures or irradiances measured or calculated for comparison with these guidelines may be averaged over a circular aperture of 1 mm diameter with a receptor having a cosine response except for comparison with the ELs for the eye in the retinal hazard spectral range between 400 nm and 1,400 nm, where measurements should be made with a detector having a 7 mm limiting aperture (pupil) with a receptor field-of-view of alpha-min (see Extended Sources section, below); and except for comparison with all ELs for wavelengths between 0.1–1 mm where the limiting aperture is 11 mm with a receptor having a cosine response. No modification of the intrabeam ("point-source") ELs is necessary, and modifications are not permitted for assumed pupil sizes less than 7 mm.

General measurement procedures

Detailed measurement procedures or calculational methods are beyond the scope of this document. Suffice to say that many potential sources of error exist in laser radiation measurement; and measurement of irradiance, radiant exposure, and radiance should not be attempted without a thorough knowledge of laser radiometry (Be78; Le76; Sl80; UN82).

Extended sources

In physiological optics it is customary to distinguish between a "point source" and an "extended source." But, in laser safety, exact geometrical definitions are not possible and small sources often fall into a "point source" category because of retinal heat flow and eye movements. The ELs for "extended sources" apply to sources that subtend a visual angle measured at the eye greater than "alpha-minimum" (alpha-min), which varies with exposure duration and wavelength band (see Table 4, Fig. 6). The ELs for "extended sources" and "point sources" are described with different units. Radiance $[W/(m^2 \cdot sr)]$ and integrated radiance $[J/(m^2 \cdot sr)]$ are used for extended sources; irradiance (W/m^2) and radiant exposure (J/m^2) are used for "point" source ELs.

For any extended source, there exists a viewing distance where the source appears to be so small that it can be considered a "point" source, and the "point" source ELs apply. The angular subtense of a circular source at this distance is termed alpha-min. The alpha-min is the linear angle subtending a visual solid angle at the eye equal to the quotient of the intrabeam "point-source" EL and "extended source" EL. The angular subtense of a source (i.e., the apparent visual angle) is the angle at the eye that is subtended by the source and should not be confused with the beam divergence of the laser.

EXPOSURE LIMITS

General

The ELs should be used as guidelines for controlling human exposure to laser radiation. They should not be regarded as thresholds of injury or as sharp demarcations between "safe" and "dangerous" exposure levels. Exposure at levels below the ELs should not result in adverse health effects. They incorporate the collective knowledge generated worldwide by scientific research and laser safety experience, and are based upon the best available published information. (See for example AC81, AC82, AN80, An79, Av78, Bo78, Ga81, Ha68, Ha78, Har78, Lu81, Ma78, MH82, Sl71, Sl80, Su82, UN82, Wo74). The ELs for ocular and skin exposure in Tables 1, 2, and 3 are to be used as given for the indicated wavelength ranges. ELs for the eye are always specified at a plane tangent to the cornea at the point of the optical axis of the eye.

Wavelength corrections

The ELs for wavelengths between 700 nm and 1,050 nm increase with increasing wave length by a factor (C_A). C_A increases from 1 to 5 as wavelength increases from 700 to 1,050 nm as shown in Fig. 1. Between 1,050 nm and 1,400 nm, the ELs include a constant spectral correction factor (C_A) of five. At exposure durations exceeding 10 seconds, a photochemically induced retinal injury is produced for short visible wavelengths. Therefore, the difference between the ocular ELs for short wavelengths (less than 550 nm) and those for longer visible wavelengths (550–700 nm) increases with greater exposure durations up to 10,000 seconds. To adjust for this changing retinal sensitivity with wavelength, another wavelength correction factor C_B is used. At wavelengths between 550 nm and 700 nm, correction factor (C_B) is applied leading to greater EL values at these longer visible wavelengths for exposure durations exceeding 10 seconds. Figure 2 shows C_B.

The ELs for skin exposure are given in Table 3. The ELs for the skin also increase by the spectral correction factor (C_A) given in Fig. 1 for wavelengths between 700 nm and 1,050 nm.

To aid in the determination of those ELs for certain exposure durations requiring calculations of fractional powers, Figs. 3, 4, and 5 may be used.

Available data are not sufficient to define additional wavelength corrections (relative to the extensive data base at 10.6 μm) over the entire IR range (1.4 μm–1 mm). At 1.54 μm, the EL given in Tables 1, 2, and 3 may be increased by a factor of 100 for exposure durations shorter than 1 μs (St81). No further extrapolation to other wavelengths is justified on the basis of presently available information.

At wavelengths greater than 1,400 nm, for beam cross-sectional areas between 100 cm^2 and 1,000 cm^2, the EL for exposure durations exceeding 10 seconds is 10,000/A_s mW/cm^2, where A_s is the area of the exposed skin in cm^2. For exposed skin areas exceeding 1000 cm^2, the EL is 10 mW/cm^2 (Ro74).

Exposure duration

Determining the applicable EL for a specific laser exposure requires a determination of the exposure duration. Some lasers operate continuously, and are termed continuous wave (CW) lasers. For a single pulse exposure, this duration is obvious; however, the following criteria should be followed where repeated exposures or lengthy exposures occur.

For any single-pulse laser exposure, the exposure duration is the pulse duration, t, defined at its half-power points. For all skin ELs and for ocular exposure to nonvisible wavelengths (less than 400 nm or greater than 700 nm), the CW exposure duration is the maximum time, T, of anticipated direct exposure. For exposure of the eye to any CW laser, the exposure duration is the maximum time of anticipated direct viewing. However, if purposeful staring into a visible (400–700 nm) beam is not intended or anticipated, then the aversion response time, 0.25 seconds, should be used. For ocular exposures in the near-infrared (700–1400 nm), a maximum exposure duration of 10 seconds provides an adequate hazard criterion for either unintended or purposeful staring conditions. In this case, eye movements will provide a natural exposure limitation and thereby eliminate the need for consideration of exposure durations greater than 10 seconds, except for unusual conditions. In special applications, such as intentional exposure from diagnostic medical instrumentation, even longer exposure durations may apply.

Exposure limits for pulse durations less than 1 ns cannot be provided at the present time because of a lack of biological data. However, a conservative interim guideline would be to limit peak irradiances to the exposure limit

TABLE 1. *Exposure limits for direct ocular exposures (intrabeam viewing) from a laser beam**

Wavelength, λ (nm)	Exposure duration, t (s)	Exposure limit (EL)	Exposure limit (EL)
180–302		3.0×10^{1} J/m^2†	3.0×10^{-3} J/cm^2†
303		4.0×10^{1} J/m^2†	4.0×10^{-3} J/cm^2†
304		6.0×10^{1} J/m^2†	6.0×10^{-3} J/cm^2†
305		1.0×10^{2} J/m^2†	1.0×10^{-2} J/cm^2†
306		1.6×10^{2} J/m^2†	1.6×10^{-2} J/cm^2†
307		2.5×10^{2} J/m^2†	2.5×10^{-2} J/cm^2†
308	10^{-9}–3×10^{4}	4.0×10^{2} J/m^2†	4.0×10^{-2} J/cm^2†
309		6.3×10^{2} J/m^2†	6.3×10^{-2} J/cm^2†
310		1.0×10^{3} J/m^2†	1.0×10^{-1} J/cm^2†
311		1.6×10^{3} J/m^2†	1.6×10^{-1} J/cm^2†
312		2.5×10^{3} J/m^2†	2.5×10^{-1} J/cm^2†
313		4.0×10^{3} J/m^2†	4.0×10^{-1} J/cm^2†
314		6.3×10^{3} J/m^2†	6.3×10^{-1} J/cm^2†
315–400	10^{-9}–10	$5.6 \times 10^{3}\ \sqrt[4]{t}$ J/m^2	$0.56\ \sqrt[4]{t}$ J/cm^2
315–400	10–10^{3}	1.0×10^{4} J/m^2	1.0 J/cm^2
315–400	10^{3}–3×10^{4}	10 W/m^2	10^{-3} W/cm^2
400–700	10^{-9}–1.8×10^{-5}	0.005 J/m^2	5×10^{-7} J/cm^2
400–700	1.8×10^{-5}–10	$18\ (t/\sqrt[4]{t})$ J/m^2	$1.8\ (t/\sqrt[4]{t}) \times 10^{-3}$ J/cm^2
400–550	10–10^{4}	100 J/m^2	1.0×10^{-2} J/cm^2
550–700	10–T_1	$18\ (t/\sqrt[4]{t})$ J/m^2	$1.8 \times 10^{-3}\ (t/\sqrt[4]{t})$ J/cm^2
550–700	T_1–10^{4}	$100\ C_B$ J/m^2	$1.0 \times 10^{-2}\ C_B$ J/cm^2
400–700	10^{4}–3×10^{4}	$0.01\ C_B$ W/m^2	$1.0 \times 10^{-6}\ C_B$ W/cm^2
700–1050	10^{-9}–1.8×10^{-5}	$0.005\ C_A$ J/m^2	$5.0 \times 10^{-7}\ C_A$ J/cm^2
700–1050	1.8×10^{-5}–10^{3}	$18\ C_A\ (t/\sqrt[4]{t})$ J/m^2	$1.8 \times 10^{-3}\ (C_A \cdot t/\sqrt[4]{t})$ J/cm^2
1050–1400	10^{-9}–5×10^{-5}	0.05 J/m^2	5.0×10^{-6} J/cm^2
1050–1400	5×10^{-5}–10^{3}	$90\ (t/\sqrt[4]{t})$ J/m^2	$9.0 \times 10^{-3}\ (t/\sqrt[4]{t})$ J/cm^2
700–1400	10^{3}–3×10^{4}	$3.2\ C_A$ W/m^2	$3.2 \times 10^{-4}\ C_A$ W/cm^2
1400–10^{6}	10^{-9}–10^{-7}	100 J/m^2	1.0×10^{-2} J/cm^2
1400–10^{6}	10^{-7}–10	$5600\ \sqrt[4]{t}$ J/m^2	$5.6 \times 10^{-1}\ (\sqrt[4]{t})$ J/cm^2
1400–10^{6}	10–3×10^{4}	1000 W/m^2	1.0×10^{-1} W/cm^2

* The limiting aperture for all ELs for wavelengths in the range 100–1000 μm is 11 mm. For all other skin ELs and for UV, IR-B and IR-C ocular ELs, the limiting aperture is 1 mm. For ocular ELs in the visible and IR-A region, the limiting aperture is 7 mm. Modification factors are:

$C_A = 1$ for $\lambda = 400$–700 nm

$C_A = 10^{[0.002(\lambda-7000)]}$ for $\lambda = 700$–$1{,}050$ nm

$C_A = 5$ for $\lambda = 1{,}050$–$1{,}400$ nm

$C_B = 1$ for $\lambda = 400$–550 nm

$C_B = 10^{[0.015(\lambda-550)]}$ for $\lambda = 550$–700 nm

$T_1 = 10$ s for $\lambda = 400$–550 nm

$T_1 = 10 \times 10^{[0.02(\lambda-550)]}$ for $\lambda = 550$–700 nm.

All values of t are in seconds.

† Not to exceed $0.56\ \sqrt[4]{t}\ \mathrm{J \cdot cm^{-2}}$ for $t \leq 10$ s.

TABLE 2. *Exposure limits for viewing of an extended laser, e.g., a diffuse reflection or a large diameter laser beam**

Wavelength, λ (nm)	Exposure duration, t (s)	Exposure limit (EL)	Exposure limit (EL)
200–400	10^{-9}–3×10^4	Same as Table 1	Same as Table 1
400–700	10^{-9}–10	$100 \sqrt[3]{t}$ kJ/(m^2·sr)	$10 \sqrt[3]{t}$ J/(cm^2·sr)
400–550	10–10^4	210 kJ/(m^2·sr)	21 J/(cm^2·sr)
550–700	10–T_1	$38.3 (t/\sqrt[4]{t})$ kJ/(m^2·sr)	$3.83\ t/\sqrt[4]{t}$ J/(cm^2·sr)
550–700	T_1–10^4	$210\ C_B$ kJ/(m^2·sr)	$21\ C_B$ J/(cm^2·sr)
400–700	10^4–3×10^4	$0.021\ C_B$ kW/(m^2·sr)	$2.1\ C_B \times 10^{-3}$ W/(cm^2·sr)
700–1400	10^{-9}–10	$100\ C_A \sqrt[3]{t}$ kJ/(m^2·sr)	$10\ C_A \sqrt[3]{t}$ J/(cm^2·sr)
700–1400	10–10^3	$38.3\ C_A (t/\sqrt[4]{t})$ kJ/(m^2·sr)	$3.83\ C_A (t/\sqrt[4]{t})$ J/(cm^2·sr)
700–1400	10^3–3×10^4	$6.4\ C_A$ kW/(m^2·sr)	$6.4 \times 10^{-1}\ C_A$ W/(cm^2·sr)
1400–10^6	10^{-9}–3×10^4	Same as Table 1	Same as Table 1

* C_A, C_B, T_1 are the same as in the footnote to Table 1.

TABLE 3. *Exposure limits for skin exposure from a laser beam**

Wavelength, λ (nm)	Exposure duration, t (s)	Exposure limit (EL)	Exposure limit (EL)
200–400	10^{-9}–3×10^4	same as Table 1	same as Table 1
400–1400	10^{-9}–1×10^{-7}	$0.2\ C_A$ kJ/m^2	$0.02\ C_A$ J/cm^2
400–1400	10^{-7}–10	$11\ C_A \sqrt[4]{t}$ kJ/m^2	$1.1\ C_A \sqrt[4]{t}$ J/cm^2
400–1400	10–3×10^4	$2\ C_A$ kW/m^2	$0.2\ C_A$ W/cm^2
1400–10^6	10^{-9}–3×10^4	same as Table 1	same as Table 1

* The limiting aperture for all skin ELs for wavelengths in the range 100–1,000 μm is 11 mm. For all other skin ELs, the limiting aperture is 1 mm. C_A is the same as in footnote to Table 1.

TABLE 4. *Values of alpha-min for two spectral bands**

Exposure duration, t (s)	400–1049 nm Angle (mrad)	1050–1400 nm Angle (mrad)
10^{-9}	8.0	11.3
10^{-8}	5.4	7.6
10^{-7}	3.7	5.2
10^{-6}	2.5	3.5
10^{-5}	1.7	2.4
10^{-4}	2.2	2.2
10^{-3}	3.6	3.6
10^{-2}	5.7	5.7
10^{-1}	9.2	9.2
1.0	15	15
10	24	24
10^2	24	24
10^3	24	24
10^4	24	24

* For all angles less than limiting angle, use intrabeam viewing ELs.

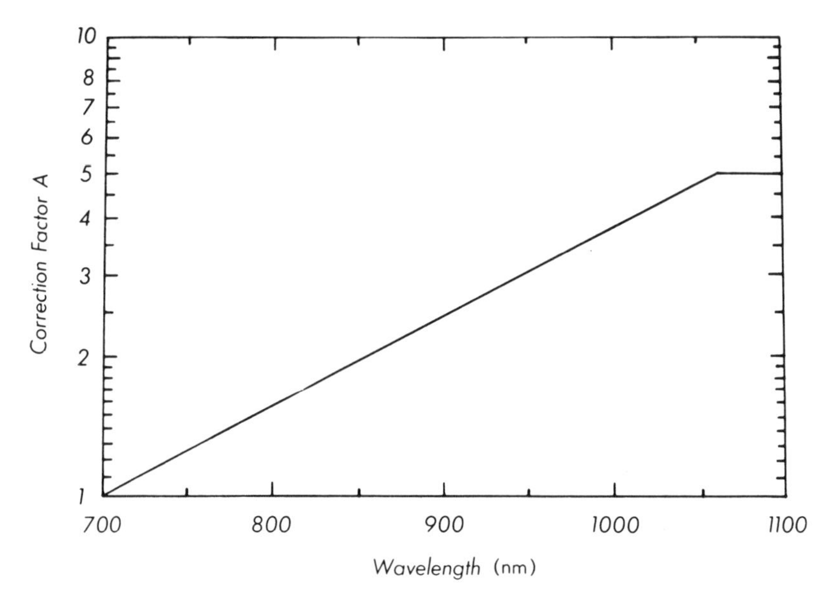

Figure 1. Exposure limit correction factor C_A for λ = 700–1,400 nm. For λ = 700–1,049 nm, $C_A = 10^{[0.002(\lambda-700)]}$. For λ = 1,050–1,400 nm, C_A = 5.

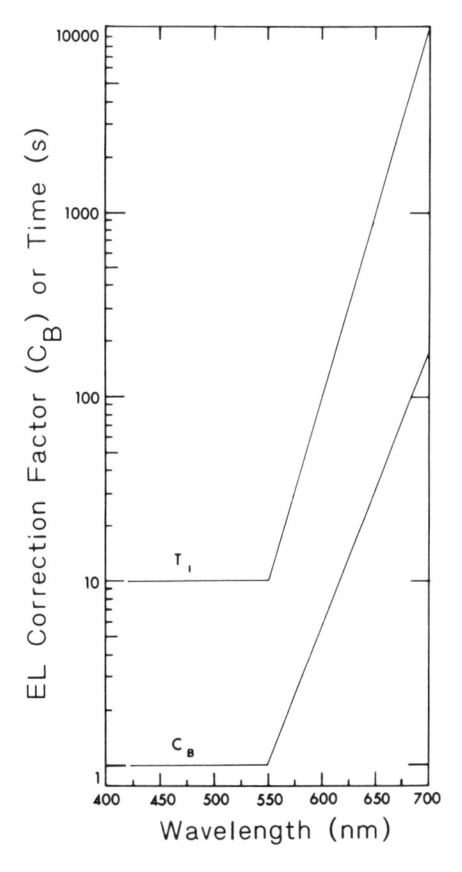

Figure 2. Exposure limit correction factor C_B for continuous wave or repetitively pulsed laser exposure of durations greater than 10 s.

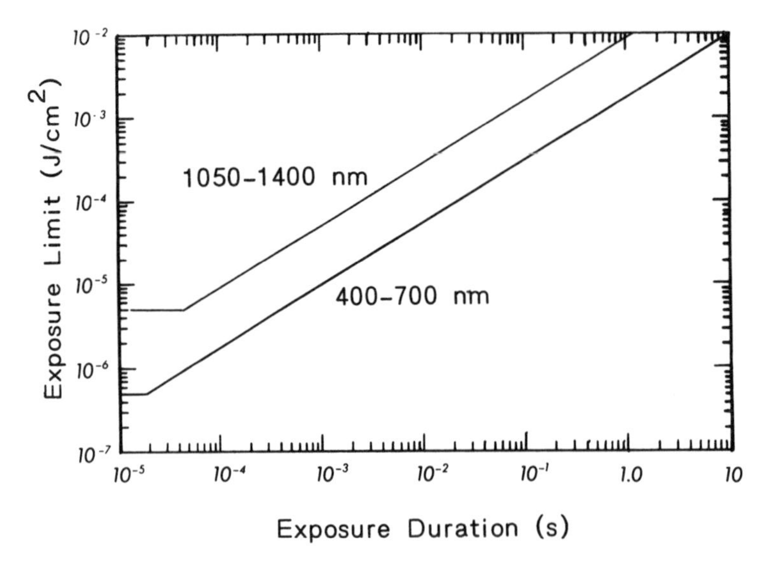

Figure 3(a). Exposure limits for intrabeam (direct) viewing of pulsed laser radiation (400–700 nm and 1,050–1,400 nm).

applicable to nanosecond pulses at the wavelength of interest.

Repetitive laser exposures

Repeated exposure within any one day to laser radiation can occur from multiple repeated exposures to a CW beam or from exposures to repetitively pulsed lasers and some scanning beam lasers. Scanning beams create repetitive-pulse exposures to the eye in the retinal hazard region (400–1,400 nm). Both the individual pulse duration and the total cu-

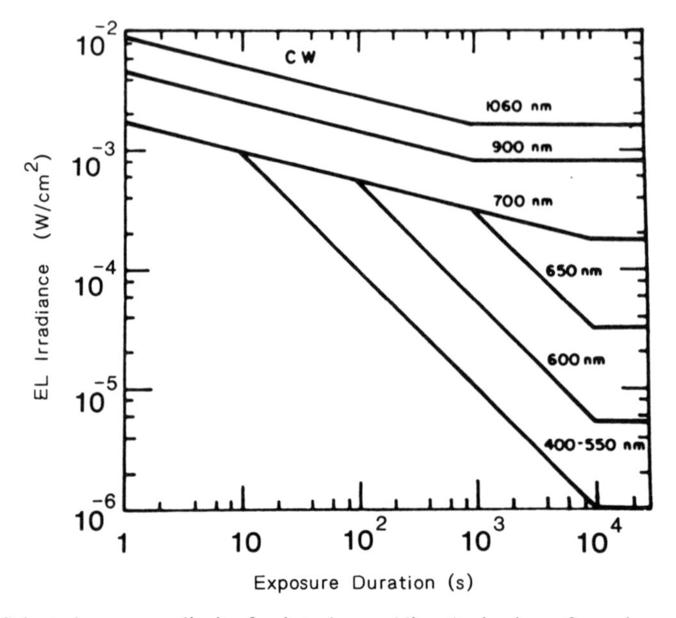

Figure 3(b). Selected exposure limits for intrabeam (direct) viewing of continuous wave laser beam (400–1,400 nm).

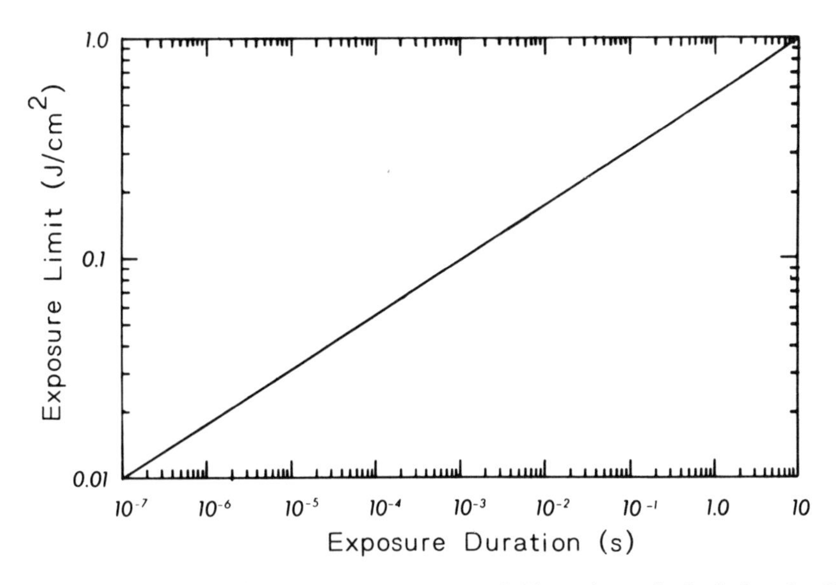

Figure 4(a). Exposure limits for pulsed laser exposure of skin and eyes for far-infrared radiation (wavelengths greater than 1.4 μm).

mulative exposure duration must be determined. In this case, the total exposure duration of the train of pulses is determined in the same manner as is used for cw laser exposures, that is, the time, T, elapsed from the beginning of the exposure (first pulse) to the end (last pulse). The methods for determining the ELs for repetitive laser exposures are as follows:

Repeated exposures, UV (315–400 nm) laser radiation. For repeated exposures, the exposure dose is additive over a 24-hour period, regardless of the repetition rate. The EL for any 24-hour period should be reduced by a factor of 2.5 times relative to the single-pulse EL if exposures on succeeding days are expected (Zu80).

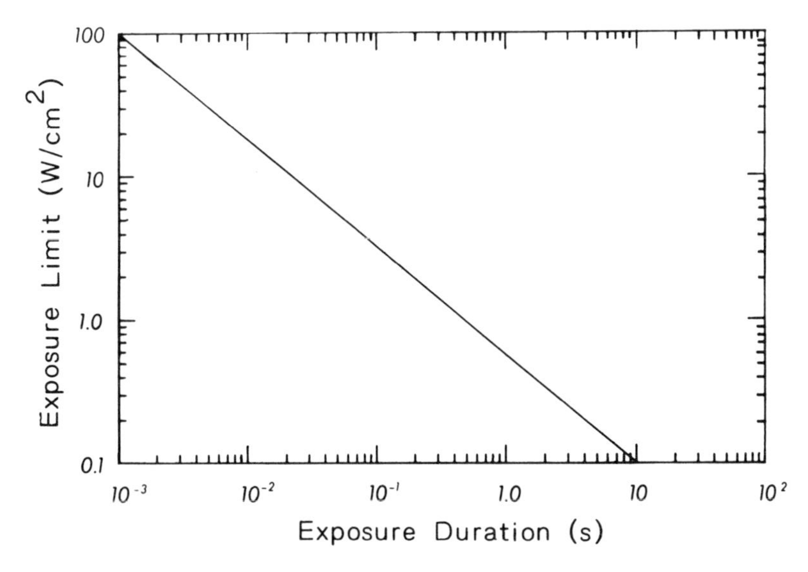

Figure 4(b). Exposure limits for continuous wave laser exposure of skin and eyes for far-infrared radiation (wavelengths greater than 1.4 μm).

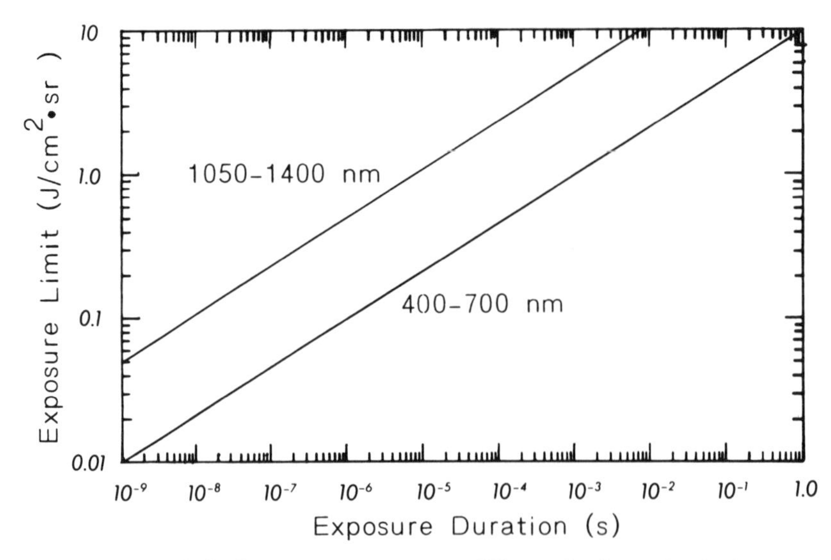

Figure 5(a). Exposure limits for extended sources or diffuse reflections of pulsed laser radiation (400–700 nm and 1,050–1,400 nm).

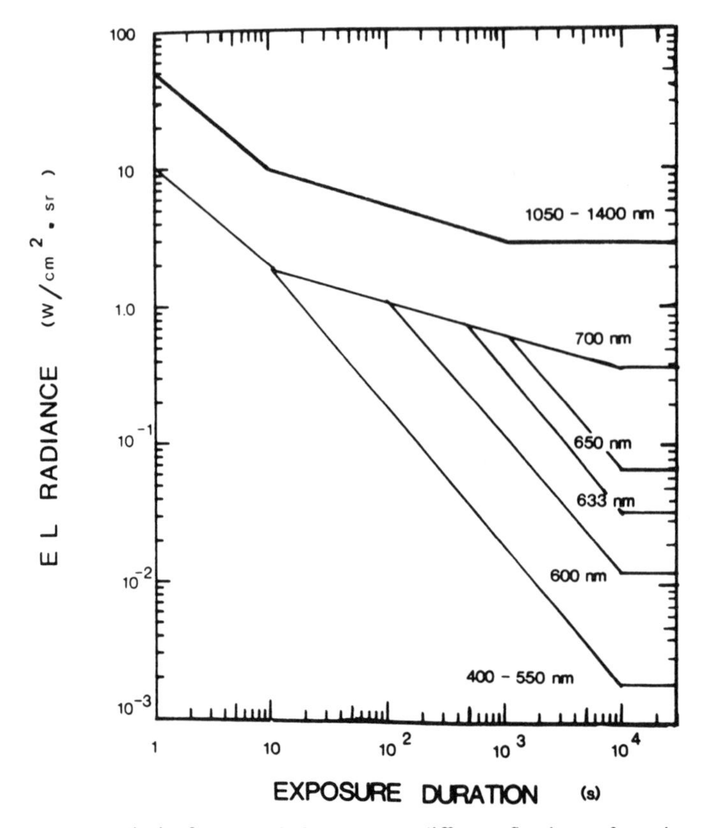

Figure 5(b). Exposure limits for extended sources or diffuse reflections of continuous wave laser radiation (400–1,400 nm).

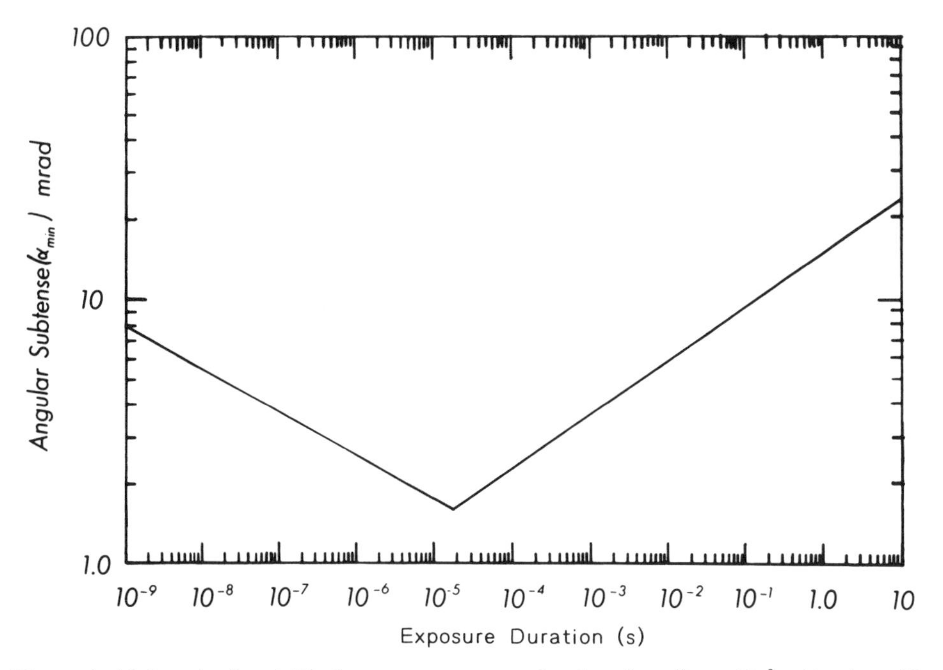

Figure 6. Alpha-min for visible laser exposure at pulse durations from 10^{-9} s (1 ns) to 10 s. Application to IR-A wavelengths for durations greater than 100 μs. The value of alpha-min for IR-A wavelengths at pulse durations less than 18 μs is 2.2 times the graph's value. Alpha-min for all exposures greater than 10 s (400–1,400 nm) is 24 mrad.

Repeated ocular exposures, visible (400–700 nm) and infrared (greater than 700 nm up to 1 mm) for both scanned CW lasers and repetitively pulsed exposure conditions. The EL per pulse for repetitively pulsed intrabeam viewing is $n^{-1/4}$ times the EL for a single pulse of the same duration (t), where n is the number of pulses found from the product of the pulse repetition frequency (prf) and the total exposure duration (T). The duration T is determined in the same manner as for a CW laser of the same wavelength (see Exposure Duration section, above). This EL applies to all wavelengths greater than 700 nm (thermal injury). For wavelengths less than 700 nm, where photochemical damage mechanisms may also apply, the EL as calculated on the basis of $n^{-1/4}$ also must not exceed the EL calculated for nt seconds when t is the duration of a single pulse in the train and nt is greater than 10 seconds. For pulse repetition frequencies greater than 15 kHz, the average irradiance or radiant exposure (radiance or integrated radiance) of the pulse train shall not exceed the EL for a CW exposure for the viewing duration T. Average irradiance is the total radiant exposure delivered during time T, divided by T.

Repeated exposure of the skin. For repetitive-pulsed laser exposure of the skin, the EL based upon a single-pulse exposure shall not be exceeded and the average irradiance of the pulse train shall not exceed the EL applicable for the total duration T of the pulse train as given in Table 3.

SPECIAL PRECAUTIONS

1. These ELs apply to the general population; however it should be recognized that some rare photosensitive individuals may react to UV laser radiation levels below these ELs. Therefore photosensitive individuals should take greater precautions to avoid exposure to UV laser radiation. Additionally, the ELs from 300 to 400 nm are not applicable to infants or aphakics.
2. These ELs are not intended to limit those laser exposures of patients required as a part of medical treatment.

PROTECTIVE MEASURES

The most effective laser hazard control is total enclosure of the laser and all beam paths. For those conditions where such containment of potentially hazardous radiation is not possible, partial beam enclosure, laser eye protectors, administrative controls, and restricted access to beam paths may be necessary. Laser safety standards have been developed worldwide (AC81; AN80; Co78; Har78; MH82; Sl80; UN82; US80), which make use of a hazard classification scheme to permit specification of control measures based upon risk posed by the laser. In some laser operations, control measures are also necessary for electrical and fire hazards, x rays, noise and airborne contaminants (these are generally only encountered with class 4 operations).

CONCLUDING REMARKS

Laser technology is still new. Because of the rapid increase in applications and use of lasers in industry, science, medicine, and in the consumer sector, the importance of laser exposure limits will increase. There is a growing need for codes of practice for all potentially hazardous laser operations.

EXAMPLES

Exposure limits for some typical lasers are provided in Appendix 1.

RATIONALE

The rationale for the exposure limits is provided in Appendix 2.

REFERENCES

AC81 American Conference of Governmental Industrial Hygienists, 1981, *A Guide for Control of Laser Hazards* (Cincinnati, OH: ACGIH).

AC82 American Conference of Governmental Industrial Hygienists, 1982, *Rationale for the Threshold Limit Values for Chemical Substances and Physical Agents in the Workroom Environment* (Cincinnati, OH: ACGIH).

AN80 American National Standards Institute, 1980, *Safe Use of Lasers. Standard Z-136.1* (New York: ANSI).

An79 Anderson F. A., 1979, *Biological Bases for and Other Aspects of a Performance Standard for Laser Products,* U.S. Food and Drug Administration, Rockville, MD, HEW Publ. (FDA) 80-8092.

Av78 Avdeev P. S., Berezin Yu. D., Gudakovskii Yu. P., Muratov V. R., Murzin A. G. and Fromzel V. A., 1978, "Experimental determination of maximum permissible exposure to laser radiation of 1.54 μ wavelength," *Sov. J. Quantum Electronics* **8,** 137–141.

Be78 Bergqvist T., Hartmann B. and Kleman B., 1978, "Imaging properties of the eye and interaction of laser radiation with matter," pp. 55–71, in: *Current Concepts in Ergophthalmology* (Edited by B. Tengroth and D. Epstein) (Stockholm: Karolinska Institute, Department of Ophthalmology).

Bi78a Birngruber R., Gabel V. P. and Hillenkamp F., 1978, "Threshold criteria and derivation of safety levels for laser radiation," pp. 73–80, in: *Current Concepts in Ergophthalmology* (Edited by B. Tengroth and D. Epstein) (Stockholm: Karolinska Institute, Department of Ophthalmology).

Bi78b Birngruber R., 1978, *Experimentelle und Theoretische Untersuchungen zur thermischen Schaedigung des Augenhintergrundes durch Laserstrahlung,* Ph.D. Dissertation, J. W. von Goethe University, Frankfurt, Federal Republic of Germany (in German).

Bo62 Boettner E. A. and Wolter J. R., 1962, "Transmission of the ocular media," *Invest. Ophthal.* **1,** 776–783 (AD 282100). See also Boettner E. A., 1967, *Spectral Transmission of the Eye,* U.S. Air Force School of Aerospace Medicine, Brooks Air Force Base, TX 78235, Final Report AF 41 (609)-2996 (July) (AD 663246).

Bo78 Borland R. G., Brennan D. H., Marshall J. and Viveash J. P., 1978, "The role of fluorescein angiography in the detection of laser-induced damage to the retina: a threshold study for Q-switched, neodymium and ruby lasers," *Exp. Eye Res.* **27(4),** 471–493.

Ca70 Carpenter J. A., Lehmiller D. J. and Tredici T. J., 1970, "U.S. Air Force permissible exposure levels for laser irradiation," *Arch. Environ. Health* **20,** 171–176.

Cl70 Clarke A. M., 1970, "Ocular hazards from lasers and other optical sources," *Crit. Rev. Environ. Control* **1(3),** 307–309.

Cl80 Cleuet A. and Mayer A., 1980, *Risques liés à l'utilisation industrielle des lasers,* pp. 207–222, Institut National de Recherche et de Sécurité, 30 rue Olivier-Noyer, 75014 Paris, France, Cahiers de Notes Documentaires, No. 99 (in French).

Co78 Court L., Chevalereaud J. P. and Santucci G., 1978, "Exposé général," pp. 367–445, in: *Effets Biologiques des Rayonnements Non Ionisants. Utilisation et risques associés (Proc. 9e Congrès international, Société française de Radioprotection)* (B.P. 72, F-92260 Fontenay-aux-Roses, France: SFRP) (in French).

Fa77 Fankhauser F., 1977, "Physical and biological effects of laser radiation," *Klin. Monatsbl. Augenheilkd* **170(2),** 219.

Ga81 Gabel V. P. and Birngruber R., 1981, "A comparative study of threshold laser lesions in the retinae of human volunteers and rabbits," *Health Phys.* **40(2),** 238–240.

Ge68 Geeraets W. J. and Berry E. R., 1968, "Ocular spectral characteristics as related to hazards from lasers and other light sources," *Amer. J. Ophthal.* **66,** 15–20.

Ge79 Gezondheidsraad (Health Council of the Netherlands), 1979, *Recommendations Concerning Acceptable Levels of Electromagnetic Radiation in the Wavelength Range from* 100 nm *to* 1 mm *(Micrometre Radiation),* Ministry of Health and Environmental Protection, Postbox 439-2260 AK Leidschendam, The Netherlands, Report 65E (March).

Gr80 Griess G. A., Blankenstein M. F. and Williford G. G., 1980, "Ocular damage thresholds for multiple-pulse laser exposures," *Health Phys.* **39,** 921–927.

Ha68 Ham W. T. Jr., Geeraets W. J., Williams R. C., Guerry D. and Mueller H. A., 1968, "Laser radiation protection," pp. 933–943, in: *Proc. 1st Int. Cong. of Radiation Protection* (Elmsford, NY: Pergamon Press).

Ha74 Ham W. T. Jr., Mueller H. A. and Goldman A., 1974, "Ocular hazard from picosecond pulses of Nd 'YAG' laser radiation," *Science* **185,** 362.

Ha76 Ham W. T. Jr., Mueller H. A. and Sliney D. H., 1976, "Retinal sensitivity to damage from short wavelength light," *Nature* **260(5547),** 153–155 (11 March).

Ha78 Ham W. T. Jr., Ruffolo J. J. Jr., Mueller H. A., Clarke A. M. and Moon M. E., 1978, "Histologic analysis of photochemical lesions produced in rhesus retina by short-wavelength light," *Invest. Ophthal. Vis. Sci.* **17(10),** 1029–1035.

Ha82 Ham W. T. Jr., Mueller H. A., Ruffolo J. J. Jr., Guerry D. III and Guerry R. K., 1982, "Action spectrum for retinal injury from near-ultraviolet radiation in the aphakic monkey," *Am. J. Ophthalmol.* **93(3),** 299–306.

Har78 Harlen F., 1978, "The development of laser codes of practice and maximum permissible exposure levels," *Ann. Occup. Hyg.* **21,** 199–221.

IE84 International Electrotechnical Commission, 1984, *Radiation Safety of Laser Products, Equipment Classification, Requirements and Users' Guide,* IEC, Geneva, Switzerland, IEC Publ. 802.

Ko78 Komarova A. A., Motzerenkov V. P., Skatskaia G. K., Chemny A. B. and Pivovarov N. N., 1978, "Action of reflected laser radiation on the eye," *Vestn. Oftal.* **1,** 46–50 (in Russian).

Le76 Le Bodo H., 1976, "La calorimétrie appliquée à la mesure des puissances et énergies lasers," *Bulletin d'Information du Bureau National de Métrologie* **24,** 12–18.

Lu81 Lund D. J., Stuck B. S. and Beatrice E. S., 1981, *Biological Research in Support of Project MILES,* Letterman Army Institute of Research, Presidio of San Francisco, San Francisco, CA 94129, Report No. 96 (July).

Ma70 Mainster M. A., White T. J., Tips J. H. and Wilson P. W., 1970, "Retinal-temperature increases produced by intense light sources," *J. Opt. Soc. Am.* **60,** 264.

Ma78 Marshall J., 1978, "Eye hazards associated with lasers," *Ann. Occup. Hyg.* **21(1),** 69–77.

MH82 Ministry of Health, 1982, *Sanitary Norms for Designing and Operating Lasers,* U.S.S.R. Ministry of Health, Moscow, No. 2392-81 (in Russian).

Ro74 Rockwell R. J. Jr. and Goldman L., 1974, *Research on Human Skin Laser Damage Thresholds,* U.S. Air Force School of Aerospace Medicine, Brooks Air Force Base, TX, Final Report, Contract F41609-72-C-0007 (June).

Ro78 Rocherolles R., 1978, "Dangers spécifiques du rayonnement cohérent. Dangers habituels liés aux fortes luminances, effets multiphotoniques et effets non linéaires," pp. 453–472, in: *Effets Biologiques des Rayonnements Non Ionisants, Utilisation et Risques Associés (Proc. 9e Congrès international, Société française de Radioprotection)* (B.P. 72, F-92260 Fontenay-aux-Roses, France: SFRP).

Sl71 Sliney D. H., 1971, "The development of laser safety criteria biological considerations," pp. 163–238, in: *Laser Applications in Medicine and Biology,* Vol. I (Edited by M. L. Wolbarsht) (New York: Plenum Press).

Sl80 Sliney D. H. and Wolbarsht M. L., 1980, *Safety Manual for Lasers and Other Optical Radiation Sources* (New York: Plenum Press).

St81 Stuck B. E., Lund D. J. and Beatrice E. S., 1981, "Ocular effects of holmium (2.06 μm) and erbium (1.54 μm) laser radiation," *Health Phys.* **40,** 835–846.

Su82 Suess M. J. (ed), 1982, Non-Ionizing Radiation Protection, World Health Organization Regional Publications, European Series No. 10, World Health Organization Regional Office for Europe, Copenhagen, Denmark.

UN82 United Nations Environment Programme/World Health Organization/International Radiation Protection Association, 1982, *Environmental Health Criteria* 23, *Lasers and Optical Radiation* (Geneva: WHO).

US80 U.S. Food and Drug Administration/Department of Health and Human Services, 1980, "Performance standards for laser products," 21 *Code of Federal Regulations* 1040.10 (Rockville, MD: U.S. FDA).

Va71 Vassiliadis A., 1971, "Ocular damage from laser radiation," pp. 125–162, in: *Laser Appli-*

cations in Medicine and Biology, Vol. I (Edited by M. L. Wolbarsht) (New York: Plenum Press).

Wo74 Wolbarsht M. L. and Sliney D. H., 1974, "The formulation of protection standards for lasers," in: *Laser Applications in Medicine and Biology,* Vol. II (Edited by M. L. Wolbarsht) (New York: Plenum Press).

Wo78 Wolbarsht M. L., 1978, "The effects of optical radiation on the anterior structures of the eye," pp. 29–46, in: *Current Concepts in Ergophthalmology* (Edited by B. Tengroth and D. Epstein) (Stockholm: Karolinska Institute, Department of Ophthalmology).

Zu76 Zuclich J. A. and Connolly J. A., 1976, "Ocular damage induced by near-ultraviolet laser radiation," *Invest. Ophthal.* **15(9),** 760–764.

Zu80 Zuclich, J. A., 1980, "Cumulative effects of near-UV induced corneal damage," *Health Phys.* **38,** 833–838.

Zw78 Zwick H. and Beatrice E., 1978, "Long-term changes in spectral sensitivity after low-level (514 nm) exposure," *Mod. Probl. Ophthal.* **19,** 319–325.

APPENDIX 1: USING THE EXPOSURE LIMIT TABLES

Example: To find the intrabeam EL for He-Ne (632.8 nm) for a 0.25-second exposure, use Table 1.1. First use the first column to find the wavelength. Choose the second 400–700 nm entry since the 0.25-second (aversion response) exposure duration falls between 1.8×10^{-5} second and 10 seconds (second column). The EL is then:

$$EL = 18(t/\sqrt[4]{t})\ J/m^2 = 18(0.25/\sqrt[4]{0.25})\ J/m^2$$
$$= 6.3\ J/m^2 = 6.3\ W \cdot s/m^2$$
$$= (6.3/0.25) = 25\ W/m^2 = 2.5\ mW/cm^2.$$

APPENDIX 2: RATIONALE FOR THE EXPOSURE LIMITS FOR LASERS

Background

The eye and skin are critical organs for laser radiation exposure. The type of effect, injury thresholds, and damage mechanisms vary significantly with wavelength. In addition, the consequences of overexposure of the eye are generally more serious than that of the skin. Consequently, safety standards have emphasized protection of the eye (AC81; AN80; IE84; Su82; UN82).

For purposes of specifying the ELs, the optical radiation spectrum has been divided by wavelength into seven spectral regions as shown in Table 2-1. These spectral regions were originally devised by the Photobiology Committee of the CIE and some dividing lines were not sharply defined (at 315 nm

TABLE 1-1. *Intrabeam exposure limits are applicable to many common continuous wave lasers for eye and skin exposure to laser radiation*

Laser type	Primary wavelength(s) (nm)	Exposure limit: Eye	Exposure limit: Skin
Helium-Cadmium Argon	441.6 488/514.5	a) 2.5 mW/cm^2 for 0.25 s b) 10 mJ/cm^2 for 10 to 10^4 s c) 1 μW/cm^2 for $t > 10^4$ s	0.2 W/cm^2 for $t >$ 10s
Helium-Neon	632.8	a) 2.5 mW/cm^2 for 0.25 s b) 10 mJ/cm^2 for 10 s c) 170 mJ/cm^2 for $t >$ 453 s d) 17 μW/cm^2 for $t > 10^4$ s	0.2 W/cm^2 for $t >$ 10 s
Krypton	647	a) 2.5 mW/cm^2 for 0.25 s b) 10 mJ/cm^2 for 10 s c) 280 mJ/cm^2 for $t >$ 871 s d) 28 μW/cm^2 for $t > 10^4$ s	0.2 W/cm^2 for $t >$ 10 s
Neodymium: YAG	1,064	1.6 mW/cm^2 for $t >$ 1000 s	1.0 W/cm^2
Gallium-Arsenide at room temp	905	0.8 mW/cm^2 for $t >$ 1000 s	0.5 W/cm^2 $t >$ 10 s
Helium-Cadmium Nitrogen	325 337.1	a) 1 J/cm^2 for 10 to 1000 s b) 1 mW/cm^2 for $t >$ 1000 s	a) 1 J/cm^2 for 10 to 1000 s b) 1 mW/cm^2 for $t >$ 1000 s
Carbon-dioxide (and other lasers 1.4 μm to 1000 μm)	10,600	0.1 W/cm^2 for $t >$ 10 s	0.1 W/cm^2 for $t >$ 10 s

and 400 nm), although for simplicity, some authorities choose a single dividing line (UN82). Sharp dividing lines are used in this document where a transition zone existed in Table 2-1.

Biological effects

Ultraviolet radiation effects. Short wavelength UV-B and UV-C radiation is absorbed within the cornea and conjunctiva, whereas UV-A radiation is absorbed largely in the lens (Bo62; Ge79; Zu76). Exposure to "actinic" UV (UV-B and UV-C) laser radiation may lead to the acute effects of erythema (reddening of the skin) and photokeratitis (corneal inflammation) and conjunctivitis. Far greater levels (typically a thousand-fold greater) of UV-A are required to produce photokeratitis and erythema by a photochemical mechanism. UV-A thermal injury to the skin or to the lens and cornea has not been demonstrated experimentally for exposure durations greater than 1 ms (UN82). The peak sensitivity for photokeratitis is believed to be around 270 nm with a decrease in the action spectrum in each direction. The peak of the erythema action spectrum varies from 200 to 300 nm depending upon the definition of the degree of severity and the time of assessment of the effect. Although in the actinic-UV region, the cornea is not substantially more sensitive to injury than untanned lightly pigmented skin, damage to the cornea is much more disabling (and painful) than injury to the skin. Repeated exposure of the cornea does not result in the development of natural protection. Tanning and thickening of the stratum corneum provide an increased natural protection for the skin. Although UV-A is absorbed more heavily in the lens, it now seems likely that excessive UV-B exposure is primarily effective in cataract formation.

Visible and IR-A radiation. In the visible and IR-A regions (400–1400 nm) the retina is primarily affected. This is due to the transparency of the ocular media and to the inherent focusing properties of the eye. The focusing properties in this spectral region render the retina much more susceptible to damage than any other part of the body. Under conditions of intrabeam viewing, the optical gain in irradiance from the cornea to retina is approximately 100,000. This applies for a point source of light. Most of the radiation that reaches the retina is absorbed by the retinal pigmented epithelium and by the underlying choroid (which supplies blood to much of the retina) (Ge68; Bi78b; Va71). The photopigments in the retina absorb only a small fraction of the incident radiation—perhaps less than 15%.

TABLE 2-1. *Regions of the optical radiation spectrum*

Region	Wavelength range
Ultraviolet	100 to 400 nm
UV-C	100 to 280 nm
UV-B	280 to 315 nm
UV-A	315 to 400 nm
Light (visible)	400 to 760 nm
Infrared	760 nm to 1 mm
IR-A	760 nm to 1.4 μm
IR-B	1.4 μm to 3.0 μm
IR-C	3.0 μm to 1 mm

Infrared radiation (IR-B and IR-C). In the IR-B and IR-C regions of the spectrum (>1.4 μm), the ocular media become opaque as a result of the strong absorption by the water component, a major constituent of all biological tissue. Thus the damage in this infrared region is primarily to the cornea, although lens damage has also been attributed to IR at wavelengths below 3 μm (IR-A and B). The IR damage mechanism appears to be thermal, at least for exposure durations greater than 1 μs. For shorter pulse durations the mechanism may be thermo-mechanical. The CO_2 laser at 10.6 μm that is now used in surgical applications exemplifies the thermal nature of laser tissue damage. In the IR-C region, as in the UV, the threshold for damage to the skin is comparable with that of the cornea. Nevertheless, the damage to the cornea will likely be of greater concern because of the adverse impact upon vision.

General rationale for the EL

Analysis of data. Extensive laser bioeffects studies have been used to establish a rationale for ELs. The derivation of laser ELs required a careful analysis of the physical and biological variables that most affected any variations in the reported laboratory biological data. The variables influencing the potential for injury in individuals exposed to laser radiation, the increase in severity of injury for suprathreshold exposure dose, the injury mechanisms, and the reversibility of injury were considered. Additionally, the accuracy of available radiometric instruments and the desire for simplicity in expressing the levels have influenced the ELs. The relative impact of each of these factors varies with wavelength and exposure duration (AC82; An79; Ca70; Cl70; Fa77; Ha68; Ha74; Har78; Lu81; Ma70; Ro74; Ro78; Sl71; Su82; Wo74; Zw78).

The safety factor. It is not possible to define a single "safety factor" between the threshold of injury and the EL. In each derivation a probit analysis was applied. An exposure value resulting in a 50-percent probability (ED-50) for an observed effect has the greatest statistical reliability and was therefore compared between laboratories. These intercomparisons showed excellent agreement when exposure conditions were the same. Light and electron microscopic findings of injury always appeared at exposure values below the ED-50 derived by ophthalmic examination. However, such microscopic injuries never appeared at exposures less than one tenth the ED-50 and usually appeared at a factor of 25–50% of the ED-50. Generally an order of magnitude factor between the ED-50 and the EL was thought to achieve an adequate margin of safety against significant or subjectively detectable acute injury. The ELs incorporate a safety factor to preclude acute injury or minor changes that could potentially lead to delayed effects.

Chronic exposure. The ELs were derived to preclude the development of delayed effects. Without an adequate understanding of the mechanisms of injury, there could be no assurance that injurious effects would not appear long after exposure to levels below the reported acute thresholds, perhaps many years following active use of such lasers. The ocular effects are by far the most important, and here delayed effects may well ensue from chronic exposure of the lens and anterior portion of the eye to UV-B and IR-A radiation (and perhaps to UV-A radiation as well). Hence there has been a substantial effort to understand the injury mechanism for each biologic effect in order to judge the possibility of delayed effects. A solid understanding of the exposure levels and conditions of man in his natural environment can aid substantially in establishing ELs for lengthy, repeated exposures. Several environmental epidemiological studies of solar retinitis, UV cataract, and skin cancer have further aided the assessment of the likelihood of delayed effects. The ELs for very long exposure durations were based upon levels found in more moderate outdoor environments where mankind has adapted with dress and social custom to limit the likelihood of delayed effects upon the eye and skin (Sl80).

Defining apertures

One of the problems in developing an EL for any optical radiation is the specification of the limiting aperture over which the level must be measured or calculated. For the skin, where no focusing takes place, one would like to use as small an aperture as possible. Unfortunately, several difficulties arise from the use of small apertures: more time is required to assess exposure; a more sensitive instrument is required; potential inaccuracies exist due to calibration problems; and calculations may be more difficult (Le76; Ro78; Sl80). A 1-mm aperture is about the smallest practical size for averaging of irradiance. For continuous exposure conditions, heat flow and scattering in the skin tend to eliminate any adverse effects of "hot spots" smaller than 1 mm (Ro74; Sl80). The same arguments hold for exposure of the cornea and conjunctiva to IR radiation of wavelengths greater than 1.4 μm and for UV radiation at wavelengths less than 400 nm. Furthermore, two factors that account for localized variations in beam irradiances—atmospherically induced "hot spots" and the mode structure in multimode lasers—seldom account for significantly higher localized beam irradiances within limited areas less than 1 mm in diameter.

In the development of laser product performance (emission) safety standards, a single aperture diameter of either 80 mm (assumed viewing by binoculars) or 7 mm has often been specified for simplicity, but considerations of special exposure conditions make such simplifications possible for those standards (US80). Another problem appears at wavelengths greater than 0.1 mm. At these far-infrared wavelengths the aperture size of 1 mm begins to create significant diffraction effects and calibration becomes a problem. Fortunately, "hot spots" must, by arguments of physical optics, be generally larger than at shorter wavelengths. For this reason aperture diameters of 1 cm or 11 mm (which has a 1 cm^2 area) are specified for wavelengths greater than 0.1 mm.

For exposure of the human eye within the retinal hazard region (400 nm to 1400 nm) the pupillary aperture "averages" the radiant exposure and a 7-mm aperture corresponding to a dark-adapted (dilated) pupil is standardized. Although the pupil may be smaller, the risk does not decrease proportionally with decreased pupillary area (Sl71); the "point" retinal image diameter increases with increasing pupil diameter. Only for extended-source viewing could an argument be made for increasing ELs for smaller pupil sizes. However, the possible use of certain medications that dilate the pupil argues against a sliding scale of ELs for bright ambient conditions.

Wavelength dependence

Injury thresholds for both the cornea and the retina vary considerably with wavelength. It is therefore necessary to consider the precision required to track the actual injury threshold variation with wavelength. It seems acceptable to adjust the ELs for different wavelengths but in a more simplistic manner than the actual biological data may indicate. Figure 2-1 provides the reciprocal of the retinal absorption relative to corneal irradiance, which is an indication of relative spectral effectiveness of different wavelengths in causing retinal injury (Bi78a; Bo62; Ge68; UN82). However, this curve does not consider the relative hazard to the lens of the eye in the near IR. The studies of Lund et al. show that the correction factor can increase greatly in the IR-A (Lu81). Because of the sudden onset of transmission of the ocular media at about 400 nm, their strong reabsorption at 1,400 nm and the shift of injury mechanisms at wavelengths from about 550 to 700 nm, some discontinuities exist at 400, 700, and 1,400 nm.

Injury mechanism

Thermal effects. Thermal injury to the retina resulting from temperature elevation in the retinal pigmented epithelium is the principal effect for exposure durations less than 10 seconds. Likewise, superficial thermal injury to the cornea and skin occurs at wavelengths greater than 1,400 nm.

Multiple wavelengths. Although ambient retinal temperature is important, it is an additive or contributing factor rather than the principal factor. When laser exposures occur at several different wavelengths and different time domains within a given interval, present theories cannot reliably predict the effects of interaction. It would be surprising if there were no interaction, and if each injury mechanism acted independently of the others. In the absence of clear data, in practice, the exposures are considered to be additive where the same tissue (e.g., the cornea) is the site of absorption for multiple wavelengths (Wo74).

Extremes of exposure duration.

1. All of the known injurious effects have a strong wavelength dependence. Wavelength is especially important in long-term exposures. However, little is well understood about the wavelength relationship in the ultra-short (<1 ns) or extremely long exposure

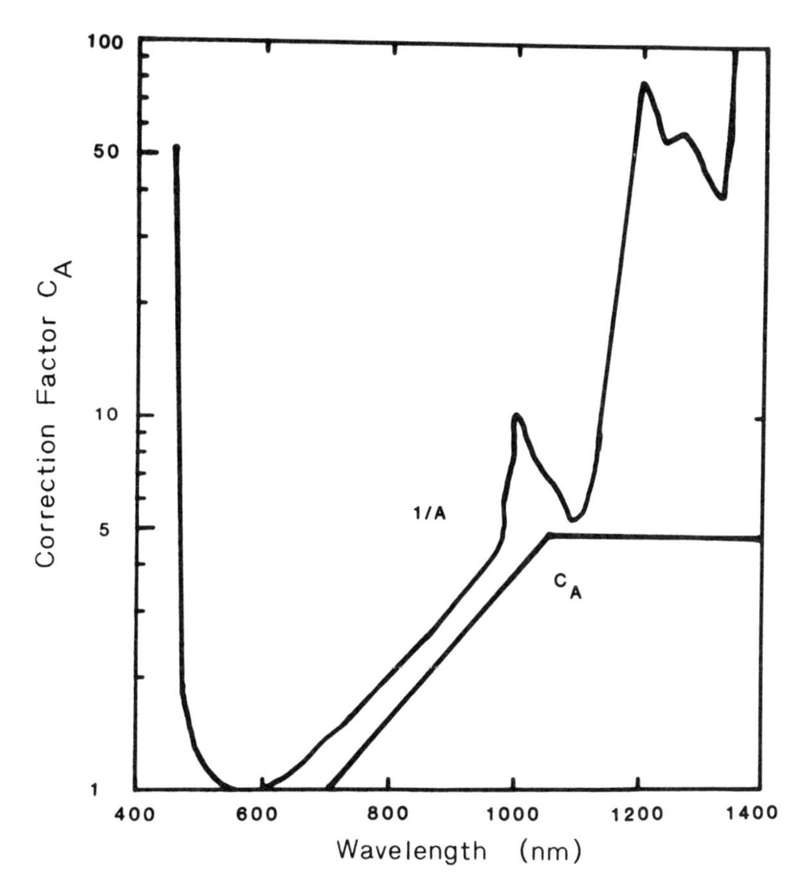

Figure 2-1. Retinal absorption and the spectral weighting factor C_A. The upper curve is the reciprocal of the retinal absorption ($1/A$) spectrum relative to corneal irradiance. It may also be thought of as the reciprocal of the action spectrum for retinal thermal injury. A more useful spectral weighting factor, C_A, is the lower function composed of straight-line segments to approximate ($1/A$) (Su82, UN82).

domains (> several hours). Further studies in these areas are underway.

2. Little data are available for repeated, long-term (chronic) exposures to laser radiation. Even studies of exposure to nonlaser sources such as bright, small-source lamps and high luminance extended sources have produced insufficient data to allow extrapolation to laser sources. Thus, the 3–8 hour exposure limits were based upon the assumption that the total retinal dose from visible and near-infrared illumination levels normally encountered in the natural environment are not hazardous (Sl80). Such an assumption required a careful study of the change of the solar spectrum with zenith angle, reflected light, etc.

3. One of the most difficult problems in developing the ELs is related to repetitive-pulse exposure when the duration of individual pulses is below 10 μs. Although extensive biological threshold data exist, the formulae for determining the EL for a train of pulses remain empirical since present theories of thermal retinal injury do not predict the pulse additivity that is actually seen (Gr80).

Infrared wavelengths. The ELs in the far-infrared region were based upon an understanding of the possible thermal effects on the cornea and a knowledge of exposures that have not resulted in adverse effects upon the eye. Because of the lack of accurate data for the infrared laser exposures of the human eye, worst-case exposure conditions were assumed. Specifically, it was assumed that absorption occurs in a very thin layer at the anterior surface of the cornea. This condition is epitomized by 10.6-micrometer CO_2 laser exposures, and fits as well exposures of the eye to any wavelength beyond approximately 3 micrometers. At wavelengths less than 3 micrometers the radiation penetrates into the cornea more deeply, and significant absorption may take place in the aqueous humor and even the lens (Av78; St81; Wo78).

The blue-light hazard. Recent studies appear to substantiate the photochemical theory that injury from chronic low-level exposure is related to absorption of short-wavelength light in the 400–520 nm region by the retinal pigmented epithelium and choroid (Ha76). In the aphakic eye UVR wavelengths greater than 300 nm reach the retina and can cause photochemical injury (Ha82). However, small temperature rises in the retina (of the order of 2 or 3 degrees) appear to be synergistic with the photochemical process so that broadband melanin pigment absorption will also play a role, albeit secondary. Shorter wavelength visible radiation has also been alleged to aggravate retinal aging (Ko78; Ma78).

Viewing conditions

For wavelengths between 400 nm and 1,400 nm (the "retinal hazard region") the ocular EL depends upon the viewing condition. For direct intrabeam viewing the retinal image may be of the order of only 10 μm in diameter, whereas for viewing diffuse laser reflections the image may be much larger (Be78; Cl80; Sl80). For this reason, there are two sets of ocular ELs in this spectral range: point-source and extended-source ELs. The point-source criteria apply to almost all intrabeam viewing situations. To aid in determining when "point-source" criteria do not apply, a limiting visual angle alpha-min (α-min) can be derived. It corresponds to the linear visual angular subtense of a circular extended source that subtends a visual solid angle equal to the ratio of the "point-source" EL in J/m^2 and the extended-source EL in $J/(m^2 \cdot sr)$. At this angle the two ELs are equivalent. Use of the "point-source" EL for subtending visual angles greater than α-min will result in an overly conservative hazard assessment. The angle α-min varies with exposure duration because of the varying degree of heat flow away from the irradiated retinal area and the degree of eye movements during the exposure.

Outside of the retinal hazard region, source size is not important and only one set of ELs for the eye is necessary; these ELs are expressed as irradiances or radiant exposures. The angle of incidence of a laser beam upon the eye and skin can be important in assessing the actual health risk, since incident rays perpendicular to the absorbing tissue are reflected the least and are therefore most dangerous. However, this factor was not specifically treated in the derivation of the ELs (Be78). The ELs were derived for worst-case absorbing conditions.

Ultraviolet radiation

Lasers that emit lines in the UV region are still relatively uncommon. UV lasers are found almost exclusively in research laboratories. The ELs for UV laser radiation are very similar to the limits for non-laser UV radiation; they are based upon the same biological data. Because of the uncertainty in the actual action spectra for photokeratitis and lenticular cataractogenesis between 300 nm and 315 nm, it was thought desirable to be slightly more conservative with the UV-B laser ELs than for non-laser source ELs. Threshold biological effect data for monochromatic UV-B laser radiation were not yet available when these guidelines were published. Some UV-A laser photokeratitis thresholds are known and were considered (Wo74; Zu76). The reader is referred to the rationale for the IRPA Guidelines on Limits of Exposure to Ultraviolet Radiation for a more detailed discussion of UV health hazards (Chapter 3).

Concluding remarks

Although laser technology is relatively new, extensive research into the bioeffects of laser radiation has taken place during the past 20 years. This research knowledge, combined with the knowledge of incoherent optical radiation effects, permits the establishment of guidelines for laser ELs (UN82). Exposure to direct laser radiation is rare since the direct beam is normally small and highly collimated, and even modest control measures limit access to the direct beam. Emphasis is normally placed upon engineering protective measures and use of eye protectors, and the ELs are only used extensively as the basis for these protective measures. Hopefully further research will expand and strengthen the foundation for these ELs. Because of the generally acute nature of laser injury, cumulative exposures to levels approaching the ELs should not be cause of concern for an increased health risk. Since the laser bioeffects of concern have thresholds, the known range of individual susceptibility is small, and considering the safety factors applied to even worst-case exposure conditions, no special limits are needed to apply to the general population. These guidelines are considered adequate for the general population as well as occupational exposure. No special assumptions such as adult ocular size, pre-exposed skin, stratum-corneum thickness or body size entered into the derivation of the limits. Only two exceptions to the foregoing need to be made; some rare photosensitive individuals or photosensitized individuals may react to UV irradiances below these ELs, and such persons should take greater precautions to avoid exposure to UV radiation. Also the limits for ocular exposure from 300 nm to 400 nm do not adequately protect the retina of an aphake or an infant (Ha82), UV absorbing lenses would be required for these individuals.

CHAPTER 5

GUIDELINES ON LIMITS OF EXPOSURE TO RADIOFREQUENCY ELECTROMAGNETIC FIELDS IN THE FREQUENCY RANGE FROM 100 kHz TO 300 GHz

THE APPLICATIONS of radiofrequency electromagnetic energy are numerous and increasing. Equipment utilizing this form of energy is to be found in industry, commerce, medicine, research, and the home. At sufficiently high intensities, exposure to radiofrequency electromagnetic fields can produce a variety of adverse health effects. Such effects include cataracts of the eye, overloading of the thermoregulatory response, thermal injury, altered behavioral patterns, convulsions, and decreased endurance. Exposure limits (ELs) are needed to protect against these adverse health effects of radiofrequency radiation exposure.

A document entitled *Environmental Health Criteria 16, Radiofrequency and Microwaves* (United Nations Environment Programme [UNEP]/World Health Organization [WHO]/ International Radiation Protection Association [IRPA], 1981) was published in 1981. The document contains a review of the biological effects reported from exposure to radiofrequency electromagnetic fields and served as the scientific rationale for the development of the IRPA/INIRC (International Non-Ionizing Radiation Committee) "Interim guidelines on limits of exposure to RF electromagnetic fields in the frequency range from 100 kHz to 300 GHz," published in 1984.

Following the publication of these guidelines in April 1984, important advances have been made in biological radiofrequency radiation research, particularly in the field of dosimetry. During its 1985 meeting, the committee concluded that the 1984 guidelines should be amended. The revised text, was approved by the IRPA Executive Council on June 2, 1987.

The IRPA Associate Societies were consulted extensively in the preparation of the original interim guidelines; a number of institutions and individuals provided comments on the current revision and their cooperation is gratefully acknowledged.

During the preparation of the present document, the composition of the IRPA/INIRC was as follows:

H. P. Jammet, Chairman (France)
J. Bernhardt (Federal Republic of Germany)
B. F. M. Bosnjakovic (Netherlands)
P. Czerski (U.S.A.)
M. Grandolfo (Italy)
D. Harder (Federal Republic of Germany)
B. Knave (Sweden)
J. Marshall (Great Britain)
M. H. Repacholi (Australia)
D. H. Sliney (U.S.A.)
J. A. J. Stolwijk (U.S.A.)
A. S. Duchêne, Scientific Secretary (France)

PURPOSE AND SCOPE

The purpose of these guidelines is to deal with the basic principles of protection against electromagnetic radiation in the radiofrequency range, so that they may serve as guidance to the various international and national bodies or individual experts who are responsible for the development of regulations, recommendations, or codes of practice to protect the workers and the general public from the potentially adverse effects of radiofrequency radiation. This document provides guidance on limits of exposure to electromagnetic radiation and fields in the frequency range from 100 kHz to 300 GHz, based on our knowledge of biological effects and on the assessment of health hazards.

The part of the frequency range 300 MHz–300 GHz is often referred to as microwave radiation (MW).

These guidelines do not apply to deliberate exposures of patients undergoing medical treatment or diagnosis.

The committee recognized that when standards on ELs are established, various value judgments are made. The validity of scientific reports has to be considered, and extrapolations from animal experiments to effects on humans have to be made. Cost vs. benefit analyses are necessary, including the economic impact of controls. The limits in these guidelines were based on the scientific data, and no consideration was given to economic impact or other nonscientific priorities. However, from presently available knowledge, the limits should provide a safe, healthy working or living environment from exposure to radiofrequency radiation under all normal conditions.

QUANTITIES AND UNITS

The physical quantities by which ELs for radiofrequency electromagnetic radiation are expressed can either refer to a situation in which a physical object is present in the area considered or a situation in which no physical object is present. In the former case, the distribution of electromagnetic energy in space will be changed by the presence of a physical object, and this is referred to as a perturbed or body-present situation. In the latter case, the electromagnetic field distribution is unperturbed, and will be referred to as the body-absent situation (Chapter 3).

In these guidelines, the *basic limits* of exposure formulated for the frequency region of 10 MHz and above are expressed by the quantity *specific absorption rate* (SAR). The specific absorption rate is the power absorbed per unit mass. The SI unit of specific absorption rate is watt per kilogram (W/kg). The SAR may be spatially averaged over the total mass of an exposed body or its parts, and may be temporally (time) averaged over a given time of exposure or over a single pulse or modulation period of the radiation. The SAR refers to the body-present situation. Radiofrequency energy absorption and methods by which the SAR can be measured or calculated for body-simulating phantoms are presented in detail in Durney et al., 1978, Durney et al., 1986, and Polk and Postow, 1986.

For practical purposes and for purposes of comparison, *derived limits* of exposure are also given in these guidelines. They are expressed in terms of power density (energy flux density) in the body-absent situation. Power density is the radiant power incident on a small sphere, divided by the cross-sectional area of that sphere. The SI unit of power density is watt per square meter (W/m^2).

In the *far field* (a plane wave) of an antenna or radiator, the electric field strength, the magnetic field strength, and the power density are related and the determination of any one of these parameters defines the remaining two.

The Poynting vector, S, is a field vector quantity equal to the vector product of the electric and magnetic field strengths, and represents the magnitude and direction of the electromagnetic power density (energy flux density):

$$S = E^2/120\pi = 120\pi H^2,$$

when S is expressed in W/m^2, E in V/m, and H in A/m.

In the *near field* or in multipath fields, both the electric and magnetic field strengths must be measured. In most cases, at frequencies below 100 MHz such measurements will be required. For the case of a complex phase relationship as found in the near field, the concept of "equivalent plane wave power density" P_{eq} (in W/m^2) has been introduced (National Council on Radiation Protection and Measurements, 1981).

For frequencies below 10 MHz, the concept of SAR has limited significance because biological effects resulting from human exposure are more fundamentally correlated with the current density generated in the body. The relationships between electric and magnetic fields outside the body and the biologically effective tissue field strengths or tissue current density have not been well developed for frequencies between 0.1 and 10 MHz. Therefore, in the frequency region below 10 MHz, basic limits are expressed in terms of the incident (outside the body) "effective electric field strength," E_{eff}, and the "effective magnetic field strength," H_{eff}. Since in the near field the phase relationship between the directional components of a field is normally unknown, the effective field strength is obtained by adding the squares of the vertical and the horizontal components and taking the square root of this sum. With relation to time, the root mean square (RMS) of each component is utilized. The SI units of E_{eff} and H_{eff} are V/m and A/m, respectively.

EXPOSURE LIMITS

Occupational

Occupational exposure to radiofrequency radiation at frequencies *below and up to 10 MHz* should not exceed the levels of unperturbed RMS electric and magnetic field strengths given in Table 1, when the squares of the electric and magnetic field strengths are averaged over any 6-minute period during the working day, provided that the body-to-ground current does not exceed 200 mA, and that any hazards of radiofrequency burns are eliminated according to the recommendations stated below (see Radiofrequency shocks and burns section). For pulsed fields, a conservative approach consists of limiting pulsed electric and magnetic field strengths as averaged over the pulse width to 32 times the values given in Table 1.

Occupational exposure to frequencies *above 10 MHz* should not exceed a SAR of 0.4 W/kg when averaged over any 6-minute period and over the whole body, provided that in the extremities (hands, wrists, feet, and ankles) 2 W per 0.1 kg shall not be exceeded and that 1 W per 0.1 kg shall not be exceeded in any other part of the body.

The limits of occupational exposure given in Table 1 for the frequencies between 10 MHz and 300,000 MHz are the working limits derived from the SAR value of 0.4 W/kg. They represent a practical approximation of the incident plane wave power density needed to produce the whole body average specific absorption rate of 0.4 W/kg, and take into account the subdivision of the radiofrequency range described in Appendix 1. These limits apply to whole body exposure from either continuous or modulated electromagnetic fields from one or more sources, averaged over any 6-minute period during the working day.

Although little information is presently available on the relation of biological effects with peak values of pulsed fields, it is suggested that the equivalent plane wave power density

TABLE 1. *Occupational exposure limits to radiofrequency (RF) electromagnetic fields*

Frequency f(MHz)	Unperturbed RMS field strength		Equivalent plane wave power density	
	Electric E(V/m)	Magnetic H(A/m)	P_{eq}(W/m^2)	P_{eq}(mW/cm^2)
0.1–1	614	$1.6/f$	—	—
>1–10	$614/f$	$1.6/f$	—	—
>10–400	61	0.16	10	1
>400–2000	$3f^{1/2}$	$0.008f^{1/2}$	$f/40$	$f/400$
>2000–300 000	137	0.36	50	5

Note: Hazards of RF burns should be eliminated by limiting currents from contact with metal objects (see Radiofrequency shocks and burns section). In most situations this may be achieved by reducing the E values from 614 to 194 V/m in the range from 0.1 to 1 MHz and from $614/f$ to $194/f^{1/2}$ in the range from >1 to 10 MHz.

as averaged over the pulse width not exceed 1,000 times the P_{eq} limits or the field strength exceed 32 times the field strength limits in Table 1 for the frequency concerned, provided that the limits of occupational exposure, averaged over any 6-minute period, are not exceeded.

For the case of the near field, where a complex phase relationship between the magnetic and electric field components exists, it is possible that exposure may be predominantly from either one of these components, and in extreme cases from the magnetic or electric field alone. The limits for magnetic and electric field strengths indicated in Table 1 for frequencies above 10 MHz may be exceeded for the case of near field exposure, provided that:

$$\tfrac{5}{6}(E^2/120\pi) + \tfrac{1}{6}(120\pi H^2) \leq P_{eq},$$

where E is the electric field strength (V/m), H is the magnetic field strength (A/m), and P_{eq} is the equivalent plane wave power density limit (W/m^2) from Table 1, and the SAR limits of occupational exposure, averaged over any 6-minute period, are not exceeded. The above formula may be applied in practical situations to the case of near field exposures in the frequency range from 10 MHz to 30 MHz, in rare instances up to 100 MHz.

Derived limits indicated in Table 1 for frequencies above 10 MHz may be occasionally exceeded provided that it can be demonstrated that the SAR resulting under particular exposure conditions as averaged over any 6-minute period and over the whole body and 0.1 kg of tissue does not exceed the basic SAR limits indicated above.

General public

Exposure of the general public to radiofrequency radiation at frequencies *below and up to 10 MHz* should not exceed the levels of unperturbed RMS electric and magnetic field strengths given in Table 2, provided that any hazards of radiofrequency burns are eliminated according to the recommendations stated below (see Radiofrequency shocks and burns section, below).

Exposures of the general public to frequencies *above 10 MHz* should not exceed a SAR of 0.08 W/kg when averaged over the whole body and over any 6-minute period.

The limits of radiofrequency exposure to the general public given in Table 2 for the frequencies between 10 MHz and 300,000 MHz are derived from the SAR value of 0.08 W/kg. They represent a practical approximation of the incident plane wave power density needed to produce the whole body average SAR of 0.08 W/kg, and take into account the subdivision of the radiofrequency range described in Appendix 1.

The limits given in Table 2 apply to whole body exposure from either continuous or modulated electromagnetic fields from one or more sources, averaged over any 6-minute period during the 24-hour day.

Although little information is presently available on the relation of biological effects with peak values of pulsed fields, it is suggested that the equivalent plane wave power density as averaged over the pulse width not exceed 1,000 times the P_{eq} limits or the field strength not exceed 32 times the field strength limits given in Table 2 for the frequency concerned, provided that the limits for public exposure,

TABLE 2. *General public exposure limits to radiofrequency electromagnetic fields*

Frequency f(MHz)	Unperturbed RMS field strength		Equivalent plane wave power density	
	Electric E(V/m)	Magnetic H(A/m)	P_{eq}(W/m^2)	P_{eq}(mW/cm^2)
0.1–1	87	$0.23/f^{1/2}$	—	—
>1–10	$87/f^{1/2}$	$0.23/f^{1/2}$	—	—
>10–400	27.5	0.073	2	0.2
>400–2000	$1.375f^{1/2}$	$0.0037f^{1/2}$	$f/200$	$f/2000$
>2000–300 000	61	0.16	10	1

averaged over 6 minutes, are not exceeded, and hazards of radiofrequency burns are eliminated.

Radiofrequency shocks and burns

Radiofrequency shocks and burns can result from touching ungrounded metal objects that have been charged up by the field or from contact of a charged up body with a grounded metal object. If the current at the point of contact exceeds 50 mA, there is a risk of burns.

The current flowing into the body has a strong dependence on the size of the object, and is a function of both the radiofrequency field and the impedance of the object to the ground. In the work environment, simple electrical measurements are sufficient to establish the risk, and precautionary measures may be instituted. In general, radiofrequency burns will not occur from currents on point contact of 50 mA or less.

Additional considerations

In view of our limited knowledge on thresholds for all biological effects, unnecessary exposure should be minimized.

Measurements of exposures to determine adherence to the guidelines should be made at positions normally occupied by persons.

The rationale for the limits is provided in Appendix 1. Measures used to protect workers and the general public from excessive or unnecessary exposure to radiofrequency radiation are given in Appendix 2.

EXPOSURE FROM MULTIPLE SOURCES

In situations where simultaneous exposure occurs from radiation emitted from sources operating at different frequencies with at least one source operating above 10 MHz, the exposure should be measured at each frequency and expressed as a fraction of the power density limit or the square of the electric or magnetic field strength limit for each frequency range in Tables 1 or 2. Then the sum of these fractions should not exceed unity.

The basic limit above 10 MHz (0.4 W/kg for occupational exposure or 0.08 W/kg for the general public) protects against thermal hazards. In multiple-source exposure situations, some contribution to the heating of specific parts of the body will result from sources operating below 10 MHz. Therefore, to provide adequate thermal protection, exposures from all sources should be combined as indicated above.

If all sources operate below 10 MHz, the exposure should be measured at each frequency and expressed as a fraction of the electric or magnetic field strength limit. Then, the sum of these fractions should not exceed unity.

EXCLUSIONS

Exposure to radiofrequency radiation emitted from low power devices, such as citizen's band radios, land mobile and marine transmitters, and walkie-talkies, can be excluded from consideration in assessing compliance with the prescribed limits provided the radiofrequency output power of the device is 7 W or less. Such devices generate only very localized fields.

EMISSION STANDARDS

Product performance or emission standards intended to provide health protection by limiting radiation emission under specified test conditions should be derived from ELs. Their numerical values may differ from these ELs because of such factors as the operating conditions and intended use of the product.

CONCLUDING REMARKS

The above guidelines will be subjected to periodic revision and amendment with advances in knowledge and identification of effects associated with particular frequencies and/or modes of generation.

REFERENCES

American National Standards Institute. American national standards safety levels with respect to human exposure to radiofrequency electro-magnetic fields, 300 kHz to 100 GHz. IEEE. New York, NY 10017, ANSI C95-1-1982; 1982.

Budinger, T. F.; Pavlicek, W.; Faul, D. D.; Guy, A. W. RF heating at 1.5 tesla and above in proton NMR imaging. Abstract from the 4th Annual Meeting of the Society of Magnetic Resonance in Medicine, 19–23 August 1985, London, United Kingdom; pp. 916–917. Available from: Society of Magnetic Resonance in Medicine, 15 Shattuck Square, #204, Berkeley, CA 94704.

Czerski, P. The development of biomedical approaches and concepts of radiofrequency radiation protection. J. Microwave Power 21:9–23; 1986.

Durney, C. H.; Johnson, C. C.; Barber, P. W.; Massoudi, H.; Iskander, M. F.; Lords, J. L.; Ryser, D. K.; Allen, S. J.; Mitchell, J. C. Radiofrequency radiation dosimetry handbook, 2nd ed. 141 p.

Brooks AFB, TX: U.S. Air Force School of Aerospace Medicine. Report No. SAM-TR-78-22; 1978.

Durney, C. H.; Massoudi, H.; Iskander, M. F. Radiofrequency radiation dosimetry handbook. 4th ed. 510 p. Brooks AFB, TX: U.S. Air Force School of Aerospace Medicine; Report No. SAM-TR-85-73; 1986.

Elder, J. A.; Cahill, D. F. Biological effects of radiofrequency radiation. United States Environmental Protection Agency, Research Triangle Park, NC. EPA-600/8-83/026F; 1984.

Foster, K. R.; Kritikos, H. N.; Schwan, H. P. Effects of surface cooling and blood flow on the microwave heating of tissues. IEEE Trans. BME-27: 313–323; 1978.

Gandhi, O. P.; Hunt, E. L.; D'Andrea, J. A. Electromagnetic power deposition in man and animals with and without ground and reflector effects. Radio Sci. 12-6S:39–47; 1977.

Gandhi, O. P. Electromagnetic radiation absorption in an inhomogeneous model of man for realistic exposure conditions. Bioelectromagnetics 3:81–90; 1982.

Gandhi, O. P.; Chatterjee, I. Radiofrequency hazards in the VLF to MF bands. Proc. IEEE. 70: 1462–1469; 1982.

Gandhi, O. P.; Chatterjee, I.; Wu, D.; Gu, Y. G. Likelihood of high rates of energy deposition in the human legs at the ANSI recommended 3–30 MHz RF safety levels. Proc. IEEE. 73:1145–47; 1985.

Gandhi, O. P. The ANSI RF safety guideline; its rationale and some of its problems. In: Dutta, S. K.; Millis, R. M., eds. Biological effects of electropollution. Philadelphia, PA: Information Ventures; 1985; pp. 9–19.

Grandolfo, M.; Michaelson, S. M.; Rindi, A. Biological effects and dosimetry of non-ionizing radiation. Radiofrequency and microwave energies. New York, NY: Plenum Press. 669 p.; 1983.

Grundler, W.; Keilmann, F.; Putterlik, V.; Strube, D. Resonant-like dependence of yeast growth rate on microwave frequencies. Brit. J. Cancer 45(supp. V):206–208; 1982.

Guy, A. W.; Chou, C.-K. Very low frequency hazard study. Final Report prepared for the USAF School of Aerospace Medicine, Brooks AFB, TX, Contract F33615-83-C-0625; 1985. Available from USAF School of Aerospace Medicine.

Illinger, K. H. Biological effects of non-ionizing radiation. ACS Symposium Series No. 157. 342 pp. Washington, DC: American Chemical Society; 1981.

ILO85 International Labour Office. International Labour Conference. Convention 161 concerning occupational health services, adopted by the International Labour Conference, Geneva, 26 June 1985. Geneva: ILO; 1985.

International Labour Office. Protection of workers against radiofrequency and microwave radiation: a technical review. Occupational Safety and Health Series No 57. 72 pp. Geneva: ILO; 1986.

Justesen, D. R. Behavioral and psychological effects of microwave radiation. Bull. N.Y. Acad. Med. 55:1058–1078; 1979.

Mumford, W. W. Heat stress due to R.F. radiation. In: Cleary, S. F., ed. Biological effects and health implications of microwave radiation. U.S. Department of Health, Education and Welfare, Rockville, MD; Publication BRH/DBE 70-2; 1970; pp. 21–34.

National Council on Radiation Protection and Measurements. Radiofrequency electromagnetic fields: properties, quantities and units, biophysical interaction and measurements. NCRP Report No. 67. 134 p. Bethesda, MD: NCRP; 1981.

National Council on Radiation Protection and Measurements. Biological effects and exposure criteria for radiofrequency electromagnetic fields, NCRP Rcport No. 86. 382 p. Bethesda, MD: NCRP; 1986.

Polk, C.; Postow, E. Handbook of biological effects of electromagnetic fields. 503 p. Boca Raton, FL: CRC Press; 1986.

Schwan, H. P. Microwave and RF hazard standard considerations. J. Microwave Power 17:1–9; 1982a.

Schwan, H. P. Nonthermal cellular effects of electromagnetic fields: AC-field induced ponderomotoric forces. Brit. J. Cancer 45(supp. V):220–224; 1982b.

Spiegel, R. J. A review of numerical models for predicting the energy deposition and resultant thermal response of humans exposed to electromagnetic fields. IEEE Trans. Microwave Theory Tech. MTT-32:730–746; 1984a.

Spiegel, R. J. Numerical modeling of thermoregulatory systems in man. In: Elder, J. A.; Cahill, D. F., eds. Biological effects of radiofrequency radiation. U.S. Environmental Protection Agency, Research Triangle Park, NC. Publication EPA-600/8-83-026F; 1984b; pp. 4-29–4-33.

Stuchly, M. A.; Kraszewski, A.; Stuchly, S. S. Exposure of human models in the near and far field—A comparison. IEEE Trans., BME-32: 609–616; 1985.

Stuchly, S. S.; Stuchly, M. A.; Kraszewski, A.; Hartsgrove, G. Energy deposition in a model of man; frequency effects. IEEE Trans., BME-33: 702–711; 1986.

Stuchly, M. A.; Spiegel, R. J.; Stuchly, S. S.; Kraszewski, A. Exposure of man in the near-field of a resonant dipole: comparison between theory and measurements. IEEE Trans., MTT-34: 26–31; 1986.

Tell, R. A.; Harlen, F. A review of selected biological effects and dosimetric data useful for development of radiofrequency safety standards for human exposure. J. Microwave Power 14:405–424; 1979.

United Nations Environment Programme/World Health Organization/International Radiation Protection Association. Environmental Health Criteria 16. Radiofrequency and microwaves. 134 p. Geneva: World Health Organization; 1981.

United Nations Environment Programme/World Health Organization/International Radiation Protection Association. Environmental Health Criteria 69. Magnetic fields. Geneva: World Health Organization; 1987.

APPENDIX 1: RATIONALE FOR EXPOSURE LIMITS

The objective of these guidelines is to protect human health from the potentially harmful effects of exposure to radiofrequency electromagnetic radiation.

Population

The first step in establishing ELs is to define the population to be protected. Exposure limits may pertain to the general population or to particular groups within it. These groups may be deemed more or less susceptible to radiofrequency-induced deleterious health effects, and may or may not be subjected to medical surveillance.

The occupationally exposed population consists of adults exposed under controlled conditions, and who are trained to be aware of potential risks and to take appropriate precautions. The duration of occupational exposure is limited to the duration of the working day or duty shift per 24 hours, and the duration of the working lifetime.

The general public comprises individuals of all ages and different health status. At higher frequencies children have a higher SAR than do adults, because of their smaller body size. The resonant frequency range is different, and so is the distribution of radiofrequency energy absorption in various body parts. Individuals or groups of particular susceptibility may be included in the general population.

In many instances members of the general public are not aware that exposure takes place, and may be unwilling to take any risks (however slight) associated with exposure. The general public can be exposed 24 hours per day, and over the whole lifetime. Finally the public cannot be reasonably expected to take precautions against radiofrequency burns and shocks.

The above considerations were the reason for adopting lower basic (and derived) ELs for the general public than for the occupationally exposed population.

Frequency ranges

Based on absorption characteristics, the radiofrequency range can be subdivided into four parts (Schwan, 1982a).

1. The subresonance range, less than 30 MHz, where surface absorption dominates for the human trunk, but not for the neck and legs, and the absorption of energy decreases rapidly with decreasing frequency.
2. The resonance range, which extends from 30 MHz to about 300 MHz for the whole body, and up to about 400 MHz if partial body (head) resonances are considered. High absorption cross sections are possible, and ELs must be set at lower values to account for worst-case situations.
3. The "hot spot" range, extending from about 400 MHz up to 2,000 MHz or even to 3,000 MHz, where significant localized energy absorption can be expected at incident power densities of about 100 W/m^2. The size of hot spots ranges from several centimeters at 915 MHz to about 1 centimeter at 3,000 MHz. Hot spots are caused by resonance or quasioptical focusing of the incident fields. The former mechanism prevails at lower frequencies, the latter at higher ones (Foster et al., 1978; Schwan, 1982a). For the human head, the hot spot range extends from 300 MHz to 2,000 MHz (Foster et al., 1978).
4. The surface absorption range.

Basic considerations

For frequencies above 10 MHz, whole body average SAR (WBA-SAR) was chosen as a quantity for establishing basic ELs, and different values were adopted for the protection of the general public and of the occupationally exposed population. The values in Tables 1 and 2 were derived primarily from the frequency dependence of WBA-SAR (Durney et al., 1978; Durney et al., 1986), and modified by considerations of the nonuniformity of radiofrequency energy deposition in various body parts, that is, of spatial peak SAR. The SAR is a convenient quantity to assess biological effects that depend on the increase in temperature associated with radiofrequency energy absorption. However, because it depends on internal electric field strength, the SAR can also be used to evaluate effects that depend on electric field strength in tissues. Therefore, although the guidelines are based primarily on thermal considerations, the objective of protection against athermal effects was also kept in mind. Another objective was to reduce hazards of radiofrequency burns and shocks.

Initially it was assumed that the local peak SAR will not exceed the WBA-SAR more than by a factor of 20, (National Council on Radiation Protection and Measurements, 1981). However, newer dosimetric data (Gandhi et al., 1985; Gandhi, 1985; Guy and Chou, 1985; Stuchly et al., 1985; M. A. Stuchly et al., 1986; S. S. Stuchly et al., 1986) indicate that under certain conditions the local SAR in the extremities, particularly in the wrists and ankles, may exceed the WBA-SAR by a factor of about 300 at certain frequencies in the resonance and subresonance ranges. Because of this an additional requirement, limiting the local SAR in extremities, was introduced in the section on occupational ELs. Although experimental data on radiofrequency energy distribution and available models do not cover the whole frequency range of interest in full detail (Elder and Cahill, 1984; Gandhi, 1985; Guy and Chou, 1985; Polk and Postow, 1986; Spiegel, 1984a; Spiegel, 1984b; M. A. Stuchly et al., 1986), the committee felt that the guidelines provide adequate protection against excessive local radiofrequency energy deposition.

Based on considerations of mechanisms of interaction, that underlie potentially harmful biological effects, both the frequency and intensity have to be taken into account. Temperature-dependent effects are well substantiated, and can be used as a basis for ELs. The emerging evidence for nonthermal mechanisms of biological effects cannot be ignored, and has to be considered in establishing ELs (American National Standards Institute, 1982; Czerski, 1986; Illinger, 1981; Schwan, 1982b).

In view of the lack of a complete understanding of the biophysics of living systems and the controversies concerning the nonthermal electromagnetic field interactions, no predictive theory based on nonthermal considerations is possible (Illinger, 1981). Empirical data must serve for establishing ELs. According to Schwan (1982b) nonthermal effects may play a significant role at frequencies below 0.1 MHz. No adequate data on biological effects at these frequencies exist, except for the extremely low frequency range, which should be considered separately (UNEP/WHO/IRPA, 1987). Therefore it seems prudent to limit the lower frequency end of the guidelines to 0.1 MHz.

In the range of 0.1–10 MHz the unperturbed magnetic field strength is less restricted than the electric field strength because it does not contribute to risk of radiofrequency burns or shock, the major reason for concern in limiting the electric field strength in the occupational exposures.

If compelling practical reasons exist (wide application at a particular frequency), a limit for a narrow range or a particular device (emission standard) operating at a frequency below 100 kHz can be set on the basis of empirical studies.

Nonthermal effects are expected to be frequency-dependent (Illinger, 1981) and may exhibit complex dose effect relationships, including intensity windows (UNEP/WHO/IRPA, 1981), both at low and high (above 30,000 MHz) frequencies (Grundler et al., 1982; Schwan, 1982b). Threshold data on adverse health effects applicable to humans over the whole frequency range and to all possible modulations do not exist. Most of the present biological data exist for frequencies in the range from 900 MHz to 10,000 MHz. Thus, where data are lacking, assumptions on possible adverse health effects must be made. These are based on our present knowledge of the biophysics of electromagnetic radiation absorption and on analytical or experimental models, as well as on limited epidemiological data. Extrapolation of limits in the range 10 MHz to 300 GHz is based on the frequency dependence of energy absorption in human beings. This may account for thermal effects but it does not allow one to predict nonthermal effects.

Energy absorption

The total amount, the distribution, and the rate of absorption of electromagnetic energy in a living system are the function of many factors (National Council on Radiation Protection and Measurements, 1981; Polk and Postow, 1986). The quantities, internal electrical field strength (V/m), induced body current (A), induced current density (A/m^2), and SAR (W/kg) are interrelated. The SAR is commonly used for comparisons of biological effects under different exposure conditions. The SAR can be calculated, or in some instances determined experimentally, for various exposure situations (see Polk and Postow, 1986).

The SAR can be used to determine the internal (absorbed) energy distribution. With some restrictions, the SAR can also be used to evaluate the time rate of change of temperature (National Council on Radiation Protection and Measurements, 1981) provided the heat exchange characteristics, including the thermoregulatory responses of the organism under consideration, are known. A more detailed discussion of SAR can be found in (National Council on Radiation Protection and Measurements, 1981; UNEP/WHO/IRPA, 1981). The SAR can be used to compute the total (integral) absorbed energy dose over the exposure time, or the specific absorption expressed in J/kg. Thus the SAR can be considered as the absorbed energy dose rate, and from the duration of exposure, the absorbed energy dose can be determined.

The SAR can be used to compare effects of exposures of unequal duration only if the duration of each of the exposures is specified. For the evaluation of effects of long-term or intermittent repeated exposures, the duration of each exposure and their spacing in time must be specified. Therefore, it follows that SAR, being a dose rate, is a more meaningful quantity if the time factor is taken into account.

The SAR is proportional to the square of the internal (*in situ*) electric field strength. The average SAR is proportional to the average of the instantaneous values of the squares of these fields. The average SAR and the SAR distribution can be computed or determined empirically by calorimetry for whole body average SAR, by thermography for average SAR and its distribution, and by implantable probes for local SARs. The peak SAR is the maximum value of a local SAR. The SAR depends on the following:

1. The incident field parameters; the frequency, intensity, polarization, and source-object configuration (far- and near-field);

2. The characteristics of the exposed body; the size, external, and internal geometry, and dielectric properties of the various tissue layers in an inhomogeneous multilayered object (such as a human or animal body);

3. Ground effects and reflector effects of other objects in the field, such as metal surfaces near the exposed body.

When the long axis of the human body is parallel to the E field vector, whole body absorption rates of electromagnetic energy approach maximal values. The amount of energy absorbed depends on a number of factors including the size of the exposed person. Standard Man (long axis 1.74 m), if not grounded, has a resonant absorption of energy at a frequency near 70 MHz. Smaller people, children, and babies experience resonant energy absorption at frequencies in excess of 100 MHz. Taller persons have a resonant absorption frequency lower than 70 MHz.

At 2,450 MHz, Standard Man will absorb about 50% of the incident electromagnetic energy. To provide some insight into the importance of the region of resonant energy absorption, it has been determined that exposure of Standard Man to 70 MHz radiation, under conditions producing maximal absorption, results in a sevenfold increase in absorption relative to that in a 2,450 MHz field. Thus values of ELs are based on the frequency dependence of human absorption over the whole frequency range for all body sizes.

In the frequency range above 10 MHz, where the working magnetic and electric field strength limits were derived from a WBA-SAR basic limit, it is possible to consider separately the contribution of the magnetic and electric fields components to the WBA-SAR. Such an analysis can be carried out using the WBA-SAR frequency dependence curves contained in the *Radiofrequency Dosimetry Handbook* (Durney et al., 1978). In the worst case, the energy coupling due to E polarization is predominantly from the electric field and the WBA-SAR reaches a maximum of about 1.2×10^{-2} W/kg at 20 MHz for a thin adult exposed at 10 W/m^2. For K polarization, the maximum possible SAR predominantly due to the magnetic field contribution for a fat person at 20 MHz is about 2×10^{-3} W/kg, that is, a factor of about six times less. The coupling of energy from the magnetic field contribution alone cannot exceed this SAR level. Therefore it is possible to modify the derived working limits for the magnetic and electric field strengths for situations where the exposure is predominantly from the magnetic or electric field components or one of these alone, provided that:

$$\tfrac{5}{6}(E^2/120\pi) + \tfrac{1}{6}(120\pi H^2) \leq P_{eq},$$

where E is the electric field strength (V/m), H is the magnetic field strength (A/m), and P_{eq} is the equivalent plane wave power density limit (W/m^2). This applies in practice to near field exposures in the frequency range between 10 MHz and 30 MHz, and in some instances up to 100 MHz.

Biological responses of exposure to radiofrequency fields do not merely depend on the intensity of the fields outside the body, but more on the fields inside the body, which are related to the average whole body SAR and its distribution within the body. Exposures to a uniform (plane-wave) electromagnetic field frequently result in a highly non-uniform deposition and distribution of energy within the body.

Only recently have methods been developed to determine the electromagnetic energy absorption in realistic exposure conditions, taking near-field exposures, ground effects, and reflector effects into account (Gandhi et al., 1977; Gandhi, 1982). Spatial average SAR, although having many limitations (UNEP/WHO/IRPA, 1981), is presently a convenient means for quantifying the relationship between biological effects and radiofrequency exposures, and for making comparisons between effects of various exposure conditions. Interspecies extrapolations must also take into account the biological differences between the species compared. Even within the same species, individual genetically determined differences influence the responses. Thermoregulatory responses differ in various species. The prediction of the magnitude of body temperature increases in different species under comparable exposure conditions and equivalent whole body SARs must take into account physiological differences, as well as the possibility of evoking general responses by localized heating of thermoregulatory centers in the hypothalamic area. Equivalent exposure parameters or SARs in animals and humans do not necessarily produce equivalent biological responses.

Health implications

A detailed review and evaluation of the scientific literature on which the ELs are based has been published (UNEP/WHO/IRPA, 1981).

The American National Standards Institute (ANSI) has produced a review of literature (ANSI, 1982) and screened only those reports that produced positive findings, were reproducible, and supplied adequate dosimetric information. Evaluations were made of these selected reports in terms of whether or not the biological effects constituted a hazard to health. Behaviour in experimental animals was found to be the most sensitive indicator of an adverse health effect (e.g., convulsion activity, work stoppage, work decrement, decreased endurance, perception of the exposing field, and aversion behaviour [Justesen, 1979]). On the basis of their review, ANSI concluded that acute (less than 1 hour) exposure to electromagnetic energy that is deposited in the whole body at an average SAR of less than 4 W/kg does not produce an adverse health effect in experimental animals. However, because prolonged exposure (days and weeks) may cause damage, a tenfold reduction in the permissible SAR (i.e., 0.4 W/kg) was invoked.

Two later comprehensive reviews (Elder and Cahill, 1984; National Council on Radiation Protection and Measurements, 1986) concluded that untoward effects can appear at lower levels, and it was suggested that the threshold WBA-SAR for health effects may be 1 W/kg (Elder and Cahill, 1984). The accumulating clinical experience with magnetic resonance imaging suggests that short-term exposure at 2 W/kg or 1 W/kg may pose problems for persons with compromised thermoregulatory function (Budinger et al., 1985). A close scrutiny of the available data revealed no need to revise the previously adopted basic WBA-SAR limits. The committee found that there is no adequate basis for identifying basic SAR limits as averaged over any gram of tissue, and instead introduced an occupational limit of 2 W per 0.1 kg to the wrists and ankles and 1 W per 0.1 kg for all other tissues. The WBA-SAR of 0.4 W/kg results in a total body EL of 28 W of absorbed energy for a 70 kg man. This limit is still excessive if concentrated in a small region, and it is therefore necessary to have an additional limit with averaging over a more limited amount of tissue. The averaging volume of 0.1 kg was chosen as a volume within which there would not be excessive gradients of energy deposition and temperature. It is necessary to choose the 100 ml volume in such a way that it is mostly within the energy absorbing region. For exposures to frequencies above 1,000 MHz it becomes necessary to describe the integrating area as a flat volume consisting of skin and subcutaneous tissue to the depth of penetration. This form of assessment allows a rational transition to the ELs for infrared radiation. Furthermore, a limitation of the body-to-ground current to 200 mA provides means of avoiding excessive heating of wrists or ankles, which may occur in special situations (Gandhi et al., 1985; Gandhi 1985; Guy and Chou, 1985). On the basis of presently available data, this constraint, and the derived ELs should be sufficient to prevent excessive radiofrequency energy deposition in any region of the body. The derived ELs suggested for public exposure should also be sufficient to provide protection in most situations normally encountered in practice.

Many observed biological effects depend on rates of energy deposition. For example, relationships have been demonstrated between the energy absorbed and temperature elevations. These relationships depend on the thermoregulatory and adaptive mechanisms under various environmental conditions. These all modify the relationship between absorbed energy and the generalized rise in body temperature. Moreover, the energy deposition is nonuniform and significant localized temperature gradients can arise making the prediction of effects difficult. For biological end points such as the cataractogenic level for the eye, a time-intensity threshold has been established. Above this threshold a total dose–response relationship has been observed. More complex relationships are evident for other biological effects where no apparent dose-response relationship exists, for example, effects on the immune system or on calcium efflux in cerebral tissue occurring within frequency or power windows (National Council on Radiation Protection and Measurements, 1986).

Threshold exposure conditions for biological effects applicable to humans exposed to all parts of the frequency range and to all possible modulation frequencies do not exist. Thus safety factors must be incorporated into the ELs to allow not only for the lack of scientific data but for all possible conditions under which the exposure might occur. Variables considered when developing the safety factors were:

1. Absorption of electromagnetic energy by humans of various sizes, with particular reference to whole or partial body resonant absorption of energy (Gandhi, 1982; Gandhi and Chatterjee, 1982; Gandhi et al., 1985; Gandhi, 1985; Grandolfo et al., 1983; Guy and Chou, 1985; M. A. Stuchly et al., 1986; S. S. Stuchly et al., 1986).
2. The lack of knowledge of the relationship between peak SAR and biological effects.
3. Environmental conditions: the ELs should be protective under adverse conditions of temperature, humidity, and air movement (Mumford, 1970; Tell and Harlen, 1979).
4. Reflection, focusing, and scattering of the incident fields in such a way that enhanced absorption occurs (Gandhi et al., 1977).
5. Possible altered response of humans taking medicines (National Council on Radiation Protection and Measurements, 1986).
6. Possible combined effects of radiofrequency electromagnetic energy with chemical or other physical agents in the environment (Czerski, 1986).
7. The possible effects of modulated microwave fields on the central nervous system and the possible existence of "power" and "frequency" windows for such effects (National Council on Radiation Protection and Measurements, 1986).
8. Possible nonthermal effects (Grandolfo et al., 1983; Grundler et al., 1982; Illinger, 1981; National Council on Radiation Protection and Measurements, 1986; Schwan, 1982a).

At present the data on 5., 6., 7., and particularly 8. are insufficient to make either a health risk assessment or even to determine if these effects present a potential health concern.

For occupational exposure a time-averaging period of 6 minutes was selected, and the same period is recommended for determining compliance with public ELs.

It was felt fundamentally important that ELs should protect against direct physical hazards as well as adverse biological effects. Although radiofrequency shocks would normally produce effects ranging from inconvenience to severe burns to tissue, situations could arise where such shocks or burns result in accident situations having more serious consequences. It is not feasible to predict or prescribe precise protection strategies against radiofrequency burns since much depends on actual configurations and conditions.

The threshold for a radiofrequency burn with point finger contact to a conducting surface is 200 mA (National Council on Radiation Protection and Measurements, 1986). Direct measurements of current between a subject and ground or an object using a simple ammeter are sufficient to verify the maximum current that can occur in a particular situation. Currents of less than 50 mA can be considered safe.

For general public exposure below 10 MHz, the committee felt that ELs should be low enough so that radiofrequency shocks cannot occur, since it would be totally unreasonable to require this group to take precautions to avoid such shocks.

The committee considered the recent data linking electric and magnetic field exposure to increased cancer risks or congenital anomalies in various exposed human populations (see UNEP/WHO/IRPA, 1987 for review). Available data are inconclusive and cannot be used for establishing ELs.

APPENDIX 2: PROTECTIVE MEASURES

The responsibility for the protection of workers and the general public against the potentially adverse effects of exposure to radiofrequency fields should be clearly assigned to a department, agency, or individual. Such responsibility at the national level should include:

1. Development and adoption of ELs and the implementation of a compliance program.
2. Development of equipment performance and emission standards limiting exposure from individual devices. These emission limits should be derived from the ELs. There should also be an evaluation of protection against emissions from existing or newly manufactured equipment and installations.
3. Radiofrequency sources may be classified according to their output and protection guides for each particular class developed. Low power sources, where emission levels cannot exceed the ELs, may form an "exempt" class, that is, such a class where no precautions are necessary.
4. Requiring emission controls to be built into mass-produced equipment to avoid dependence for safety on instructions and labels alone. Safe design of equipment allows the use of additional protective devices to be reduced to a minimum.
5. Drafting of guidelines or codes of practice for users or operators on the safe use of radiofrequency electromagnetic energy.
6. Development of standardized measurement procedures and survey techniques.
7. Educating workers and the general public on the valid concerns about exposure to radiofrequency radiation and the measures being taken to protect them.

At the operational level, the following procedures should be implemented:

1. Surveys should be conducted in all installations or on all devices likely to emit radiofrequency radiation above accepted limits. Many sources of radiofrequency radiation are unconfined emitters (radiobroadcasting, TV, radars, and similar) and their radiations propagate over large areas. Appropriate siting and a careful analysis of the health and environmental impact are necessary before a site is selected. Following installation, analytical estimates of the distribution of electromagnetic fields should be verified by empirical measurements. For proper siting, it is necessary to determine the use and occupancy of surrounding areas. Zones that are clearly marked with appropriate warning signs may be useful.
2. Survey reports should be retained and include details of the exposure conditions, and in cases where ELs are exceeded, indications of ways and means for reducing levels. Recommendations for reduction of exposures to acceptable levels should be implemented as soon as possible.
3. Require that all radiofrequency radiation workers familiarize themselves with the local safety procedures or codes of safe practice.
4. Ensure that operational procedures are such that the use of personnel protective devices (special eye/head protectors, clothing) is a last resort.

A more complex description of measures for the protection of radiofrequency workers may be found in International Labour Office (1986). There is no specific medical surveillance for radiofrequency workers and thus general rules established by the ILO in the Occupational Health Services Convention 1985 (No. 161) apply (ILO85).

CHAPTER 6

INTERIM GUIDELINES ON LIMITS OF EXPOSURE TO 50/60 Hz ELECTRIC AND MAGNETIC FIELDS

INTRODUCTION

JUST OVER 100 years ago, human exposure to external electric and magnetic fields was limited to those fields arising naturally. Within the past 50 years there has been significant growth of man-made, extremely low frequency (ELF) electromagnetic fields at frequencies of 50 and 60 Hz predominantly from electric energy generation, transmission, distribution, and use. Man-made ELF fields are now many orders of magnitude greater than the natural fields at 50 and 60 Hz.

Within all organisms are endogenous electric fields and currents that play a role in the complex mechanisms of physiological control, such as neuromuscular activity; glandular secretion; cell-membrane function; and development, growth, and repair of tissue. It is not surprising that, because of the role of electric fields and currents in so many basic physiological processes (Grandolfo et al., 1985), questions arise concerning possible effects of artificially produced fields on biological systems. With advances in technology and the ever greater need for electric energy, human exposure to 50/60 Hz electric and magnetic fields has increased to the point that valid questions are raised concerning safe limits of such exposure.

Public concern is growing, and in many countries regulatory and advisory agencies have been requested to evaluate possible adverse effects of ELF electromagnetic fields on human health (Grandolfo and Vecchia, 1989). From a review of the scientific literature, it is apparent that gaps exist in our knowledge, and more data need to be collected to answer unresolved questions concerning biological effects of exposure to these fields. On the other hand, analysis of the existing literature does not provide evidence that exposure at present day levels has a public health impact that would require corrective action. In several countries there was and still is an ongoing controversy between proponents of restrictive protective measures and advocates of technological growth leading to an increase in exposure levels. It thus appeared that there was a need for guidelines on exposure limits (ELs) based on an objective analysis of currently available knowledge. A detailed discussion of potential adverse effects can be found in the literature (Ahlbom et al., 1987; United Nations Environment Programme [UNEP]/World Health Organization [WHO]/International Radiation Protection Association [IRPA], 1984; UNEP/WHO/IRPA, 1987), and a summary is presented in the Rationale for Exposure Limits section.

The International Radiation Protection Association/International Non-Ionizing Radiation Committee (IRPA/INIRC) recognizes that when ELs are established, various value judgments have to be made. The validity of scientific reports must be considered, and extrapolation from animal experiments to effects on humans has to be made. A cost–benefit analysis, taking into account national public health priorities, and considerations of economic impact and social issues, may be necessary to derive limits suited to the conditions prevailing in different countries.

The rationale for these interim guidelines is provided in the Rationale for Exposure Limits section. Measures used to protect workers and the general public from excessive or unnecessary exposure to 50/60 Hz fields are given in the section on Protective Measures.

A first draft of these interim guidelines was distributed to the Associate Societies of IRPA and to various institutions and individual scientists for comments. Many helpful comments and criticisms were obtained and are gratefully acknowledged.

During the preparation of these guidelines, the composition of the IRPA/INIRC was as follows:

H. P. Jammet, Chairman (France)
J. H. Bernhardt (Federal Republic of Germany)
B. F. M. Bosnjakovic (The Netherlands)
P. Czerski (U.S.A.)
M. Grandolfo (Italy)
D. Harder (Federal Republic of Germany)
B. Knave (Sweden)
J. Marshall (Great Britain)
M. H. Repacholi (Australia)
D. H. Sliney (U.S.A.)
J. A. J. Stolwijk (U.S.A.)
A. S. Duchêne, Scientific Secretary (France).

These interim guidelines were approved by the President of IRPA on behalf of the IRPA Executive Council on May 3, 1989.

PURPOSE AND SCOPE

These guidelines apply to human exposure to electric and magnetic fields at frequencies of 50 or 60 Hz. The guidelines do not apply to deliberate exposure of patients undergoing medical diagnosis or treatment.

QUANTITIES AND UNITS

Transmission lines and electrical devices generate 50/60 Hz electric and magnetic fields in their vicinity. The electric and magnetic fields must be considered separately because at the very long wavelengths (thousands of kilometers in free space or air) corresponding to these frequencies, measurements are made in the near field of the source where the electric and magnetic fields are not in a constant relationship. Biological systems are extremely small compared to these wavelengths so the electric and magnetic fields interact (couple) separately with the system.

The electric field created in the vicinity of a charged conductor is a vector quantified by the electric field strength, E. This vector is the force exerted by an electric field on a unit charge and is measured in volts per meter (V m^{-1}). The E-vector either oscillates along a fixed axis (single-phase source) or rotates in a plane and describes an ellipse (three-phase source). Because the electric field at or close to the surface of an object in the field is generally strongly perturbed, the value of the "unperturbed electric field" (i.e., the field that would exist if all objects were removed) is used to characterize exposure conditions.

The magnetic field is a vector quantity. As in the case of electric fields, single-phase and three-phase fields can be defined whose vector properties are the same as those previously described for the E-field. The magnetic field strength, H, is the axial vector whose curl (rotation) equals the current density vector, including the displacement current, and is expressed in amperes per meter (A m^{-1}). The magnetic flux density, B, also known as the magnetic induction or simply the B-field, is accepted, however, as the most relevant quantity for expressing magnetic fields associated with biological effects. The magnetic flux density is defined in terms of the force exerted on a charge moving in the field and has the unit tesla (T). One tesla is equal to 1 V s m^{-2} or 1 weber per square meter (Wb m^{-2}). An important distinction between B- and H-fields becomes apparent only in a medium that has a net polarization of magnetic dipoles. In free space, and for practical purposes in biological tissues, B and H are proportional. The ratio B:H is the magnetic permeability of free space, $\mu_0 = 4\pi\ 10^{-7}$ H m^{-1}, and it is expressed in henrys per meter (1 H m^{-1} = 1 Wb A^{-1} m^{-1}).

The E-, B-, and H-fields can be described each as having time-varying sinusoidal components along three orthogonal axes. The effective field strength is the root of the sum of these three mean squared (temporal mean square) mutually orthogonal components.

In this document, ELs for the magnetic field are given in terms of the root mean square (RMS) magnetic flux density. The corresponding values of the RMS magnetic field strength can be obtained taking into account that 1 μT corresponds to 0.7958 A m^{-1}, and 1 A m^{-1} corresponds to 1.257 μT.

The quantities described above characterize somewhat idealized exposure conditions (fields impinging upon the surface of the body) because reference is made to the situation in which the exposed body is absent from the field. Thus, unperturbed E or H fields may be compared to radiometric quantities.

Biological effects should be related to the field on the surface of the body, as well as to the electric fields, currents, and current densities induced inside the body. The unit of electric current is the ampere (A), which is equal to an electric charge of 1 coulomb moving past a given point per second (C s^{-1}). The current density is a vector quantity whose magnitude is equal to the charge that crosses a unit surface area perpendicular to the flow of charge per unit of time. The current density is expressed in amperes per square meter (A m^{-2}). These quantities should be considered dosimetric. Considered rigorously, these quantities represent dose rates. In order to derive a meaningful dose concept, the dependence of biological effects upon the duration of exposure and the distribution of the dose rate in space and time have to be explored and taken into account.

Well-established effects, such as interactions with excitable membranes of nerve and muscle cells, show a dependence upon local E field strength or current density. As is the case with other dose-rate-dependent phenomena, thresholds for these effects can be demonstrated. These thresholds are best expressed in terms of the current density induced in the body. Thus, the criterion used for ELs is this induced current density. Because currents induced in the body cannot be easily measured directly, the working limits in terms of unperturbed electric field strength and magnetic flux density have been derived from the criterion value of induced current density. The values obtained were modified taking into account effects due to indirect coupling mechanisms as discussed in the rationale.

A review of quantities, units, and terminology for non-ionizing radiation protection has been previously published (IRPA/INIRC, 1985).

EXPOSURE LIMITS

The basic criterion is to limit current densities induced in the head and trunk by continuous exposure to 50/60 Hz electric and magnetic fields to no more than about 10 mA m^{-2}.

Occupational

Electric field. Continuous occupational exposure during the working day should be limited to RMS unperturbed electric field strengths not greater than 10 kV m^{-1}.

Short-term occupational exposure to RMS electric field strengths between 10 and 30 kV m^{-1} is permitted, provided the RMS electric field strength (kV m^{-1}) times the duration of exposure (hours) does not exceed 80 for the whole working day.

Magnetic field. Continuous occupational exposure during the working day should be limited to RMS magnetic flux densities not greater than 0.5 mT.

Short-term occupational whole-body exposure for up to 2 hours per workday should not exceed a magnetic flux density of 5 mT. When restricted to the limbs, exposures up to 25 mT can be permitted.

General public

Electric field. Members of the general public should not be exposed on a continuous basis to unperturbed RMS electric field strengths exceeding 5 kV m^{-1}. This restriction applies to open spaces in which members of the general public might reasonably be expected to spend a substantial part of the day, such as recreational areas, meeting grounds, and the like.

Exposure to fields between 5 and 10 kV m^{-1} should be limited to a few hours per day. When necessary, exposures to fields in excess of 10 kV m^{-1} can be allowed for a few minutes per day, provided the induced current density does not exceed 2 mA m^{-2} and precautions are taken to prevent hazardous indirect coupling effects.

It should be noted that buildings in a 5-kV m^{-1} external field have a field strength lower by more than an order of magnitude inside the building.

Magnetic field. Members of the general public should not be exposed on a continuous basis to unperturbed RMS magnetic flux densities exceeding 0.1 mT. This restriction applies to areas in which members of the general public might reasonably be expected to spend a substantial part of the day.

Exposures to magnetic flux densities between 0.1 and 1.0 mT (RMS) should be limited to a few hours per day. When necessary, exposures to magnetic flux densities in excess of 1 mT should be limited to a few minutes per day.

Summary of exposure limits

A summary of the limits recommended for occupational and general public exposures to 50/60 Hz electric and magnetic fields is given in Table 1.

MEASUREMENT

Measurements of electric and magnetic fields should be performed according to the IEC and IEEE standards on measurement of electric and magnetic fields from AC power lines (International Electrotechnical Commission, 1987; Institute of Electrical and Electronics Engineers, 1987). For inhomogeneous magnetic fields, the magnetic flux density should be averaged on a loop surface of 100 cm^2.

PROTECTIVE MEASURES

The responsibilities for the protection of workers and the general public against the potentially adverse effects of exposure to 50/60 Hz electric and magnetic fields should be clearly assigned. It is recommended that the competent authorities consider the following steps:

- development and adoption of ELs and the implementation of a compliance program;
- development of technical standards to reduce the susceptibility to electromagnetic interference, for example, for pacemakers;
- development of standards defining zones with limited access around sources of strong electric and magnetic fields because of electromagnetic interference (e.g., for pacemakers and other implanted devices). The use of appropriate warning signs should be considered;
- requirement of specific assignment of responsibility for the safety of workers and the public to a person at each site with high exposure potentials;
- drafting of guidelines or codes of practice

TABLE 1. *Limits of exposure to 50/60 Hz electric and magnetic fields*

Exposure characteristics	Electric field strength $kV\ m^{-1}$ (RMS)	Magnetic flux density mT (RMS)
Occupational		
Whole working day	10	0.5
Short term	30[a]	5[b]
For limbs	—	25
General public		
Up to 24 $h\ d^{-1}$ [c]	5	0.1
Few hours per day[d]	10	1

[a] The duration of exposure to fields between 10 and 30 $kV\ m^{-1}$ may be calculated from the formula $t \leq 80/E$, where t is the duration in hours per work day and E is the electric field strength in $kV\ m^{-1}$.

[b] Maximum exposure duration is 2 h per work day.

[c] This restriction applies to open spaces in which members of the general public might reasonably be expected to spend a substantial part of the day, such as recreational areas, meeting grounds, and the like.

[d] These values can be exceeded for a few minutes per day provided precautions are taken to prevent indirect coupling effects.

for worker safety in 50/60 Hz electromagnetic fields;
- development of standardized measurement procedures and survey techniques;
- requirements for the education of workers on the effects of exposure to 50/60 Hz fields and the measures and rules that are designed to protect them.

General rules on medical surveillance have been established by the International Labour Office in the ILO Convention 161 concerning Occupational Health Services (International Labour Office, 1985).

CONCLUDING REMARKS

The ELs are based on established or predicted effects of exposure to 50/60 Hz fields. Although some epidemiological studies suggest an association between exposure to 50/60 Hz fields and cancer, others do not. Not only is this association not proven, but present data do not provide any basis for health risk assessment useful for the development of ELs.

Current laboratory studies are testing the hypothesis that 50/60 Hz fields may act as, or with, a cancer promoter. These studies are still exploratory in nature and have not established any human health risk from exposure to these fields.

These limits have been developed from present knowledge, but there are still areas of research where questions have been raised that need to be addressed. A major research effort to supplement our knowledge on the health consequences, if any, of long-term continuous exposure of humans to low-level 50/60 Hz fields is required.

There is an ever-increasing number of people wearing implanted cardiac pacemakers that may be sensitive to interference from electric and magnetic fields. These people may not always be adequately protected against interference at some of the above ELs (see the section on Cardiac pacemakers).

These guidelines will be subjected to periodic revision and amendment with advances in knowledge.

RATIONALE FOR EXPOSURE LIMITS

General considerations

These guidelines are intended to protect the health of humans from the potentially harmful effects of exposure to electric and magnetic fields at frequencies of 50/60 Hz and are primarily based on established or predicted effects.

Population. The first step in establishing ELs is to define the population to be protected. Exposure limits may pertain to the general population or to particular groups within it.

A distinction is made between the ELs for workers and the general public for the following reasons. The occupationally exposed population consists of adults exposed under controlled conditions in the course of their duties who should be trained to be aware of potential risks and to take appropriate precautions. Occupational exposure is limited to the duration of the working day or duty shift per 24 hours and the duration of the working lifetime.

The general public comprises individuals of all ages and different health status. Individuals or groups of particular susceptibility may be included in the general population. In many instances, members of the general public are not aware that exposure takes place or may be unwilling to take any risks (however slight) associated with exposure. The general public can be exposed 24 hours per day and over a whole lifetime. Finally, the public cannot be expected to accept effects such as annoyance and pain due to transient discharges or hazards due to contact currents. The above considerations were the reason for adopting lower ELs for the general public than for the occupationally exposed population.

Coupling mechanisms. The more important mechanisms of these interactions (Bernhardt, 1988; Tenforde and Kaune, 1987) are as follows:

- electric fields (50/60 Hz) induce a surface charge on an exposed body that results in currents inside the body, the magnitude of which is related to the surface charge density. Depending on the exposure conditions, size, shape, and position of the exposed body in the field, the surface charge density can vary greatly resulting in a variable and nonuniform distribution of currents inside the body;
- magnetic fields from 50/60 Hz sources also act on humans by inducing electric fields and currents inside the body;
- electric charges induced in a conducting object (e.g., an automobile) exposed to a

50/60 Hz electric field may cause current to pass through a human in contact with it;
- magnetic field coupling to a conductor (e.g., a wire fence) causes 50/60 Hz electric currents to pass through the body of a person in contact with it;
- transient discharges (often called sparks) can occur when people and metal objects exposed to a strong electric field come into sufficiently close proximity;
- electric or magnetic fields (50/60 Hz) may interfere with implanted medical devices (e.g., unipolar cardiac pacemakers) and cause malfunction of the device.

The first two interactions listed above are examples of direct coupling between living organisms and 50/60 Hz fields. The latter four interactions are examples of indirect coupling mechanisms because they can occur only when the exposed organism is in the vicinity of other bodies. These bodies can include other humans or animals, and objects such as automobiles, fences, or implanted devices.

Criterion for limiting exposure

The limits recommended in these guidelines were developed primarily on established or predicted immediate health effects produced by currents induced in the body by external electric and magnetic fields. These limits correspond to induced current densities that are generally at or slightly above those normally occurring in the body (up to about 10 mA m^{-2}).

An unperturbed electric field strength of 10 kV m^{-1} induces RMS current densities of less than 4 mA m^{-2} when averaged over the head or trunk region (Bernhardt, 1985; Kaune and Forsythe, 1985). However, peak current densities in the same regions would exceed 4 mA m^{-2} (Dimbylow, 1987; Kaune and Forsythe, 1985) depending on the size, posture, or orientation of the person in the electric field.

Assuming a 10-cm radius loop of tissue of conductivity 0.2 S m^{-1}, a magnetic flux density of 0.5 mT at 50/60 Hz would induce an RMS current density of about 1 mA m^{-2} at the periphery of the loop.

The following statements can be made with respect to induced current density ranges and biological effects resulting from whole-body exposure to 50/60 Hz fields (UNEP/WHO/IRPA, 1987):

1. Between 1 and 10 mA m^{-2}: minor biological effects have been reported;
2. Between 10 and 100 mA m^{-2}: there are well-established effects, including visual and nervous system effects;
3. Between 100 and 1,000 mA m^{-2}: stimulation of excitable tissue is observed, and there are possible health hazards;
4. Above 1,000 mA m^{-2}: extra systoles and ventricular fibrillation can occur (acute health hazards).

Endogenous current densities in the body are typically up to about 10 mA m^{-2}, although they can be much higher during certain functions. The committee felt that, to be conservative, current densities induced by external electric or magnetic fields should not significantly exceed this value. Thus, limits for continuous human exposure to electric and magnetic fields were determined using this criterion.

Safety factors in health protection standards do not guarantee safety but represent an attempt to compensate for unknowns and uncertainties. Readers are referred to the Environmental Protection Agency (1986) for a description of the use of safety factors in the derivation of ELs.

Rationale for limits on electric field exposures

From a review of laboratory and human studies, the conclusions below were drawn by a joint WHO/IRPA Task Group studying health effects of ELF electric fields (UNEP/WHO/IRPA, 1984). The guidelines are essentially based on the following WHO/IRPA conclusions and on more recent reports:

1. Animal experimentation indicates that exposure to strong ELF electric fields can alter cellular, physiological, and behavioral events. Although it is not possible to extrapolate these findings to human beings at this time, these studies serve as a warning that unnecessary exposure to strong electric fields should be avoided.
2. Adverse human health effects from exposure to ELF electric fields at strengths normally encountered in the environment or the workplace have not been established.
3. The threshold field strength for some human beings to feel spark discharges in electric fields is about 3 kV m^{-1}, and to perceive the field is between 2–10 kV m^{-1}. There are

no scientific data at this time that suggest that perception of a field per se produces a pathological effect.

4. Although there are limitations in the epidemiological studies that suggest an increased incidence of cancer among children and adults exposed to 50/60 Hz fields, the data cannot be dismissed. Additional study will be required before these data can serve as a basis for risk assessment.

5. It is not possible from present knowledge to make a definitive statement about the safety or hazard associated with long-term exposure to sinusoidal electric fields in the range of 1 to 10 kV m^{-1}. In the absence of specific evidence of particular risks or disease syndromes associated with such exposure, and in view of experimental findings on the biological effects of exposure, it would be prudent to limit exposure, particularly for members of the general population.

Basis for extrapolation of experimental results to man. External electric fields induce electric currents within biological systems. The magnitude of the induced currents depends on a number of factors, including the size and shape of the object exposed, its electric conductivity, and proximity to other conducting objects. Man's size and posture make it difficult to simulate in laboratory animals the current densities that occur when man is exposed to strong electric fields. The species differences between man and laboratory animals may result in differences in the threshold for biological responses, the magnitude of physiological responses, and the degree of adaptation.

A physical basis for extrapolations, or what is called "scaling" from animal to human subjects, was provided by recent dosimetric studies. Comparing enhancement of fields at body surfaces and internal current densities, comparisons of exposure can be made. According to a study by Kaune et al. (1985), exposure of pigs to an effective electric field strength of 25 kV m^{-1} is equivalent to human exposure at 9.3 kV m^{-1} if peak electric field strengths at the surface of the body are taken into account, and 13 kV m^{-1} if the average electric field strength at the surface is considered. Using average total current densities in the torso as a scaling factor, Kaune and Forsythe (1988) derived approximate values for comparisons of exposure of humans, swine, and rats. Electric fields at 60 Hz result in current densities 7.3 times larger in humans than in swine, and 12.5 times larger in humans than in rats at the same unperturbed field strength. Exposure of rats at 100 kV m^{-1} is roughly equivalent to human exposure at 8 kV m^{-1}, and exposure of swine at 100 kV m^{-1} to human exposure at 13.7 kV m^{-1}. Thus animal experiments suggest that prolonged exposure to fields in the range of 8–15 kV m^{-1} does not lead to evident adverse effects in humans (Czerski, 1988).

Experimental studies. A large body of data has been collected on blood chemistry changes in animals exposed under different conditions; no consistent picture of physiological changes is evident.

Results of behavioral experiments on animals, which suggested an effect of exposure, were at levels at or above those needed for sensory perception of the field. Most behavioral tests showed no effects with exposure to electric field strengths up to 10 kV m^{-1} (Ahlbom et al., 1987; UNEP/WHO/IRPA, 1984). Effects on behavior have been reported in isolated instances from electric field exposure inducing current densities as low as 3 mA m^{-2}. Health consequences, if any, of these observations require further studies.

Many studies on laboratory animals (rodents) have indicated that there are no significant adverse effects on growth and development. Multigeneration studies in swine and rats exposed to electric fields (30 kV m^{-1} and 65 kV m^{-1}, respectively) revealed developmental defects (Phillips, 1981; Phillips, 1985). These results were not confirmed in recent, well-controlled studies on rats (Rommereim et al., 1988; Sikov et al., 1987).

Evaluation of the evidence from many studies indicates that animal morbidity and mortality are unaffected by long-term exposure. Such studies were carried out on small laboratory animals (rats and mice) at unperturbed 50/60 Hz electric field strengths up to 100 kV m^{-1} (Bonnell et al., 1986) and on larger animals, including miniature pigs, at levels near 30 kV m^{-1} (Phillips, 1981, 1983).

Human studies. At 50/60 Hz, a field strength of 20 kV m^{-1} is the perception threshold of 50% of people for sensation from their head hair or tingling between body and clothes. As shown under laboratory conditions, a small percentage of people can perceive a field strength of 2 or 3 kV m^{-1} (Cabanes and Gary, 1981; IEEE, 1978).

Controlled laboratory studies on volunteers exposed for short periods to electric field strengths of up to 20 kV m^{-1} have, in general, shown no significant effects (Hauf, 1974; Hauf and Wiesinger, 1973; Rupilius, 1976; Sander et al., 1982). These data do not establish that health effects could not occur from long-term exposure (months or years).

Well-controlled studies on the health status of linemen and switchyard workers have not revealed any statistically significant differences between exposed and control groups (Baroncelli et al., 1986; Knave et al., 1979; Stopps and Janischewsky, 1979). These studies are among the more complete and are representative of high levels of occupational exposure. Because of the small populations studied and the resulting low statistical power, these studies cannot exclude the existence of small effects in these highly exposed populations.

Several studies of the incidence of cancer or mortality from cancer among arbitrarily defined occupational groups considered to be exposed to electromagnetic fields (among other factors) suggested an association between "electrical occupations" and cancer. Because of the inherent uncertainty associated with this type of epidemiological study and the lack of measurement of exposure, no definitive conclusion can be drawn. However, the questions raised by these reports necessitate further investigation (Repacholi, 1988; UNEP/WHO/IRPA, 1984, 1987).

Recent epidemiological studies (Savitz et al., 1988) provided some support for the findings of a previous study on childhood cancer and exposure to weak magnetic fields (Wertheimer and Leeper, 1979). Both studies were carried out in the same geographical area and on a similar population; thus, the conclusions drawn from both reports cannot be generalized. A scientific panel (Ahlbom et al., 1987) that evaluated the implications of these epidemiological studies concluded that the association between cancer incidence and 60 Hz field exposure is still not established and remains a hypothesis. The committee concurs with this conclusion. To date, chronic low-level exposure to 50/60 Hz fields has not been established to increase the risk of cancer.

From the experimental data and human studies, it was concluded (UNEP/WHO/IRPA, 1984) that no adverse health effects resulted from short-term exposures at strengths up to 20 kV m^{-1} at frequencies of 50 and 60 Hz.

Steady-state 50/60 Hz current from contact with charged objects can produce biological effects that range from just noticeable perception to ventricular fibrillation and death (UNEP/WHO/IRPA, 1984). The severity of an electric shock from touching a charged object depends upon a number of factors, including grounding conditions, the magnitude of contact current, the duration of current flow, and body mass. Currents above the 10-mA level represent a serious risk because the "let-go" threshold* may be exceeded, and the individual might not be able to release a charged object due to involuntary muscle contractions (IEEE, 1978, 1984). The estimated level of let-go current in small children is approximately one-half of that for an adult man. If the current is increased beyond the let-go value, there is a possibility that ventricular fibrillation can occur. Short-circuit currents resulting from touching charged objects can be related to unperturbed field strengths (Guy, 1985).

Typical threshold values resulting from steady-state contact currents of 50/60 Hz from vehicles (IEEE, 1978; UNEP/WHO/IRPA, 1984; Zaffanella and Deno, 1978) include:

- 10–12 kV m^{-1}: Median pain perception for children, finger contact, car;
- 8–10 kV m^{-1}: Painful shock for children, finger contact, truck;
- 4–5 kV m^{-1}: Median touch perception for men, finger contact, car;
- 2–2.5 kV m^{-1}: Median touch perception for children, finger contact, car.

Transient capacitative discharges can occur between a person and a charged object via a spark through an air gap. The human reaction to transient electric shocks from spark discharges has been shown to depend in a complex manner on the discharge voltage and the capacitance of the discharging object (IEEE, 1978). The sensitivity of individuals to transient discharges has a linear dependence on

* The let-go threshold is the current intensity above which a person cannot let go of a gripped conductor as long as the stimulus persists due to uncontrollable muscle contraction.

body mass (Larkin et al., 1986). Other factors such as sex, age, or skin hardness have no correlation with the threshold sensitivity of an individual to transient electric discharges. Data obtained on adults exposed to spark discharges of various intensities showed that 50% of the subjects perceived spark discharges in a field of 2.7 kV m^{-1}, and 50% found the spark discharges annoying at 7 kV m^{-1} (Zaffanella and Deno, 1978). To obtain these data, persons standing in an electric field touched a metallic post with a finger; it is assumed that their capacitance was of the order of 170 pF.

Derivation of ELs. The proposed criterion of induced current density of 10 mA m^{-2} in the body is within the range of magnitude of spontaneous endogenous current densities. Our knowledge about the possible effects of long-term exposures to fields inducing currents near the criterion value is still limited, and most evidence is based on short-term observations.

In view of these reservations, the continuous occupational exposure should be limited to 10 kV m^{-1}, inducing a current density of 4 mA m^{-2} on average. There is substantial workplace experience, in addition to controlled laboratory studies on volunteers, that indicate that short-term exposures to fields up to 30 kV m^{-1} have no significant adverse health consequences. Exposures to electric fields between 10 and 30 kV m^{-1} produce proportionally increasing discomfort and stress and should be limited in duration accordingly. A practical approach to limiting the duration of exposure to fields between 10 and 30 kV m^{-1} is to use the formula $t \leqslant 80/E$ over the whole working day, where t is the duration of exposure in hours to a field strength of E kV m^{-1}.

For the reasons given in the Population section, a further safety factor was incorporated for exposure of the general public. A safety factor of five with respect to the criterion of 10 mA m^{-2} was introduced, leading to a limit of 2 mA m^{-2} that corresponds to an electric field strength of 5 kV m^{-1}.

The limit of 5 kV m^{-1} for continuous exposure of the general public also provides substantial protection from annoyance caused by steady-state contact currents or transient discharges. This limit, however, cannot completely eliminate perception of the electric field effects, since the perception threshold for some people is below 5 kV m^{-1}. In such cases, additional technical measures (e.g., grounding) may be instituted to avoid indirect coupling effects arising from touching charged, ungrounded objects. It should be noted that continuous exposures of the general public outdoors rarely exceed 1–2 kV m^{-1} (Tenforde and Kaune, 1987).

Rationale for limits on magnetic field exposures

In terms of a health risk assessment, it is difficult to correlate precisely the internal tissue current densities with the external magnetic flux density. Assuming a 10-cm radius loop in tissue of conductivity 0.2 S m^{-1}, it is possible to calculate the magnetic flux density that would produce potentially hazardous current densities in tissues. The following statements can be made for induced current density ranges and magnetic flux densities of sinusoidal homogeneous fields that produce biological effects from whole-body exposure (UNEP/WHO/IRPA, 1987):

1. Between 1 and 10 mA m^{-2} (induced by magnetic flux densities above 0.5 and up to 5 mT at 50/60 Hz)—minor biological effects have been reported.
2. Between 10 and 100 mA m^{-2} (above 5 and up to 50 mT at 50/60 Hz)—there are well-established effects, including visual and nervous system effects.
3. Between 100 and 1,000 mA m^{-2} (above 50 and up to 500 mT at 50/60 Hz)—stimulation of excitable tissue is observed and there are possible health hazards.
4. Above 1,000 mA m^{-2} (greater than 500 mT at 50/60 Hz)—extra systoles and ventricular fibrillation can occur (acute health hazards).

Several laboratory studies have been conducted on human subjects exposed to sinusoidally time-varying magnetic fields with frequencies of 50/60 Hz. None of these investigations has revealed adverse clinical or physiological changes. The strongest magnetic flux density used in these studies with human volunteers was a 5-mT, 50 Hz field to which subjects were exposed for 4 hours.

Some epidemiological reports present data indicative of an increase in the incidence of cancer among children, adults, and occupational groups. The studies suggest an association with exposure to weak 50- or 60-Hz magnetic fields. These associations cannot be satisfactorily explained by the available theoretical basis for the interaction of 50/60 Hz electromagnetic fields with living systems. The magnetic flux densities in some epidemiological studies suggesting an increased cancer incidence are at values near 0.25 μT. This magnetic flux density would induce a current density that is well below those levels normally occurring in the body. The epidemiological studies are not conclusive. Although these epidemiological data cannot be dismissed, there must be additional studies before they can serve as a basis for health hazard assessment. Furthermore, scant laboratory evidence is available to support the hypothesis that there is an association between 50/60 Hz fields and increased cancer risk.

The total number of direct observations of the effect of magnetic flux densities in humans is limited. Controlled laboratory studies on human volunteers exposed for 4 to 6 hours per day for several days to magnetic flux densities up to 5 mT (together with electric fields up to 20 kV m^{-1}) did not demonstrate significant effects (Sander et al., 1982; UNEP/WHO/IRPA, 1987). Therefore the short-term occupational exposure should not exceed 5 mT (inducing current densities of 10 mA m^{-2}, the criterion value) and 25 mT for the extremities. The latter value takes into account the loop diameters in the limbs that are about one-fifth of those in the trunk. Because of the sparseness of data on long-term exposures to magnetic fields, the magnetic flux density for continuous exposure in the occupational environment is limited to 0.5 mT, a limitation that can be accepted without great difficulty in most occupational environments.

For reasons developed earlier, the limit for continuous exposure of the general public was set at 0.1 mT, a factor of five below the limit for continuous occupational exposure and the short-term exposure limit was set at 1 mT.

Typical office and household average levels are 0.01–1 μT (Gauger 1984). Values of up to 12 μT may occur intermittently in rooms heated using electric/oil heaters (Krause 1986) as well as peak levels of 1–30 μT at a 30-cm distance from various appliances; magnetic flux densities from power transmission systems are somewhat higher and can typically approach levels of about 10–30 μT (Bernhardt, 1988; Tenforde and Kaune, 1987; UNEP/WHO/IRPA, 1987). However, near (3.0 cm) some appliances like electric blankets, hair dryers, shavers, and magnetic mains voltage stabilizers, the magnetic flux density can approach levels of 0.1–1 mT. Because of the strong inhomogeneity of magnetic fields near most appliances, the magnetic flux density should be averaged on a loop surface of 100 cm^2 to simulate a realistic current loop in the human body.

Cardiac pacemakers

Interference of electric fields with implanted cardiac pacemakers can lead to reversion to a fixed rate; cessation of stimulation is possible. Such direct interference has not been reported in fields below 2.5 kV m^{-1} (Moss and Carstensen, 1985; UNEP/WHO/IRPA 1984). Although body currents produced by contact with a vehicle in a weaker field may cause interference, the risk of pacemaker reversion is believed to be slight (UNEP/WHO/IRPA, 1984).

The probability that a malfunction will occur in the presence of an external magnetic field is strongly dependent on the pacemaker model, the value of the programmed sensing voltage, and the area of the pacemaker loop, which is determined during implantation. Assuming sensitivities of 0.5 to 2 mV for 50/60 Hz and worst-case conditions (600 cm^2 for the area of the pacemaker electrode, homogeneous field perpendicular to this area), interference magnetic flux densities of 15–60 μT may be calculated. Similar results were obtained by other authors (Bridges and Frazier, 1979). For more realistic conditions, due to the inhomogeneity of magnetic fields, smaller effective loop areas, and smaller sensitivities of the signal-sensing circuit, there is only a small probability of the occurrence of a pacemaker malfunction at magnetic flux densities below about 100–200 μT (UNEP/WHO/IRPA, 1987).

Increased sophistication of pacemakers has made the question of possible electromagnetic interference more difficult. Physicians implanting (and/or programming) very sensitive unipolar-demand pacemakers should be informed by the manufacturer that malfunction of the pacemaker can occur in a strong electric field so the patient can receive a detailed warning, for example, avoiding areas with strong electric fields. A reduction of the susceptibility of pacemakers to electromagnetic interference is recommended.

REFERENCES

Ahlbom, A.; Albert, E. N.; Fraser-Smith, A. C.; Grodzinsky, A. J.; Marron, M. T.; Martin, A. O.; Persinger, M. A.; Shelanski, M. L.; Wolpow, E. R. Biological effects of power line fields. In: New York State Power Lines Project, Scientific Advisory Panel Final Report. New York: New York State; 1987:67–87.

Baroncelli, P.; Battisti, S.; Checcucci, A.; Comba, P.; Grandolfo, M.; Serio, A.; Vecchia, P. A survey on the health conditions among workers of the Italian state railways high voltage substations. Am. J. Occup. Med. 10:45–55; 1986.

Bernhardt, J. H. The establishment of frequency dependent limits for electric and magnetic fields and evaluation of indirect effects. Radiat. Environ. Biophys. 27:1–27; 1988.

Bernhardt, J. H. Evaluation of human exposure to low frequency fields. In: The impact of proposed radiofrequency radiation standards on military operations, proceedings of a NATO Workshop. 92200 Neuilly-sur-Seine, France: AGARD, 7 rue Ancelle; AGARD-LS-138; 1985: p. 8.1–8.18.

Bonnell, J. A.; Maddock, B. J.; Male, J. C.; Norris, W. T.; Cabanes, J.; Gary, C.; Conti, R.; Nicolini, P.; Margonato, V.; Veicsteinas, A.; Cerretelli, P. Research on biological effects of power frequency fields. Proceedings of the International Conference on Large High-Voltage Electric Systems; 27 August to 4 September 1986; Paris; 1986: CIGRE (paper 36-08).

Bridges, J. E.; Frazier, M. J. The effect of 60 hertz electric and magnetic fields on implanted cardiac pacemakers. Palo Alto, CA: Electric Power Research Institute; EPRI-EA 1174; 1979.

Cabanes, J.; Gary, C. Direct perception of the electric field. In: Proceedings of the International Conference on Large High-Voltage Electric Systems, Stockholm; 1981, Paris: CIGRE.

Czerski, P. Extremely low frequency (ELF) electric fields, biological effects and health risk assessment. In: Repacholi, M. H., ed. Non-ionizing radiations. Physical characteristics, biological effects and health hazards assessment. Proceedings of an INIR Workshop, Melbourne, Australia; 1988: 255–271. Available from: Australian Radiation Laboratory, Yallambie, Victoria, Australia 3085.

Dimbylow, P. J. Finite difference calculations of current densities in a homogeneous model of a man exposed to extremely low frequency electric fields. Bioelectromagnetics 8:355–375; 1987.

Environmental Protection Agency. Federal radiation protection guidance. Proposed alternatives for controlling public exposure to radiofrequency radiation. Notice of proposed recommendations. Federal Register Vol. 51, No. 146:1986.

Gauger, J. R. Household appliance magnetic field survey. Arlington, VA: Naval Electronic Systems Command; IIT Research Institute Report EO 6549-43; 1984.

Grandolfo, M.; Michaelson, S. M.; Rindi, A., eds. Biological effects and dosimetry of static and ELF electromagnetic fields. New York and London: Plenum Press; 1985.

Grandolfo, M.; Vecchia, P. Existing safety standards for high voltage transmission lines. In: Franceschetti, G.; Gandhi, O. P.; Grandolfo, M., eds. Electromagnetic biointeraction: Mechanisms, safety standards, protection guides. New York and London: Plenum Press; 1989.

Guy, A. W. Hazards of VLF electromagnetic fields. In: The impact of proposed radiofrequency radiation standards on military operations. Proceedings of a NATO Workshop, 92200 Neuilly-sur-Seine, France: AGARD, 7 rue Ancelle: AGARD-LS-138; 1985: p. 9.1–9.20.

Hauf, R. Effects of 50 Hz alternating fields on man. Electrotechn. Z. B. 26:318–320; 1974 (in German).

Hauf, R.; Wiesinger, J. Biological effects of technical electric and electromagnetic VLF fields. Int. J. Biometeorol. 17:213–215; 1973.

Institute of Electrical and Electronics Engineers: Working Group on Electrostatic and Electromagnetic Effects. Electric and magnetic field coupling from high voltage AC power transmission lines—classification of short-term effects on people. IEEE Trans. Power Appar. Syst. 97:2243–2252; 1978.

Institute of Electrical and Electronics Engineers: Power Engineering Society Transmission and Distribution Committee. Corona and field effects of AC overhead transmission lines. Piscataway, NJ: IEEE; 1984.

Institute of Electrical and Electronics Engineers: IEEE standard procedures for measurements of electric and magnetic fields from AC power lines. New York: IEEE; ANSI/IEEE Std 644; 1987.

International Electrotechnical Commission/International Standard IEC 833. Measurement of power frequencies electric field. 1st ed. IEC-42 (Central Office) 37 (Draft); 1987.

International Labour Office. International Labour Conference. Convention 161 concerning occupational health services, adopted by the International Labour Conference, Geneva, 26 June 1985. Geneva: ILO; 1985.

Kaune, W. T.; Forsythe, W. C. Current densities measured in human models exposed to 60 Hz electric fields. Bioelectromagnetics 6:13–22; 1985.

Kaune, W. T.; Forsythe, W. C. Current densities induced in swine and rat models by power-frequency electric fields. Bioelectromagnetics 9:1–24; 1988.

Kaune, W. T.; Phillips, R. D.; Anderson, L. E. Biological studies of swine exposed to 60 Hz electric fields. Palo Alto, CA: Electric Power Research Institute; EPRI-EA 4318. Project 799.1; 1985.

Knave, B.; Gamberale, F.; Bergstrom, S.; Birke, E.; Iregren, A.; Kolmodin-Hedman, B.; Wenneberg, A. Long-term exposure to electric fields. A cross-sectional epidemiological investigation on occupationally exposed workers in high-voltage substations. Scan. J. Work Environ. Health 5: 115–125; 1979.

Krause, N. Exposure of people to static and time variable magnetic fields in technology, medicine, research, and public life: Dosimetric aspects. In: Bernhardt, J. H., ed. Biological effects of static and extremely low frequency magnetic fields. Munich: MMV Medizin Verlag; 1986: 57–71.

Larkin, W. D.; Reilly, J. P.; Kittler, L. B. Individual difference in sensitivity to transient electrocutaneous stimulation. IEEE Trans. Biomed. Eng. 33: 495; 1986.

Moss, A. J.; Carstensen, E. Evaluation of the effects of electric fields on implanted cardiac pacemakers. Palo Alto, CA: Electric Power Research Institute; EPRI-EA 3917; 1985.

Phillips, R. D. Biological effects of 60 Hz electric fields on small and large animals. In: Biological effects of static and low frequency electromagnetic fields. Proceedings of the US/USSR Scientific Exchange Program on Physical Factors Symposium; 4–8 May 1981; Kiev, USSR; 1981 (in Russian).

Phillips, R. D. Biological effects of electrical fields on miniature pigs. Proceedings of the Fourth Workshop of the US/USSR Scientific Exchange Program on Physical Factors in the Environment; 21–24 June 1983; Research Triangle Park, NC: National Institute of Environmental Health Sciences; 1985.

Repacholi, M. H. Carcinogenic potential of extremely low frequency fields. In: Repacholi, M. H., ed. Non-ionizing radiations. Physical characteristics, biological effects and health hazards assessment. Proceedings of an INIR Workshop, Melbourne, Australia; 1988:303–315. Available from: Australian Radiation Laboratory, Yallambie, Victoria, Australia 3085.

Rommereim, D. N.; Rommereim, R. L.; Anderson, L. E.; Sikov, M. R. Reproductive and teratologic evaluation in rats chronically exposed at multiple strengths of 60 Hz electric fields. Abstracts of the 10th Annual Meeting of the Bioelectromagnetics Society; 19–23 June 1988; Stamford, CT. Gaithersburg, MD: The Bioelectromagnetics Society; 1988: 37.

Rupilius, J. P. Investigations on the effects on man of an electrical and magnetic 50 Hz alternating field. Freiburg, Germany: Albert Ludwig University; 1976 (in German). Dissertation.

Sander, R.; Brinkman, J.; Kuhne, B. Laboratory studies on animals and human beings exposed to 50 Hz electric and magnetic fields. Proceedings of the International Conference on Large High-Voltage Electric Systems; 1–9 September 1982; Paris: CIGRE, (Paper 36-01); 1982.

Savitz, D. A.; Wachtel, H.; Barnes, F. A.; John, E. M.; Tvrdik, J. G. Case-control study of childhood cancer and exposure to 60-Hz magnetic fields. Am. J. Epidemiol. 128:21–38; 1988.

Sikov, M. R.; Rommereim, D. N.; Beamer, J. L.; Buschbom, R. L.; Kaune, W. T.; Phillips, R. D. Developmental studies of Hanford miniature swine exposed to 60-Hz electric fields. Bioelectromagnetics 8:229–242; 1987.

Stopps, G. J.; Janischewsky, W. Epidemiological study of workers maintaining HV equipment and transmission lines in Ontario. Vancouver, BC, Canada: Canadian Electrical Association Research Report; 1979.

Tenforde, T. S.; Kaune, W. T. Interaction of extremely low frequency electric and magnetic fields with humans. Health Phys. 53:585–606; 1987.

United Nations Environment Programme/World Health Organization/International Radiation Protection Association. Environmental health criteria 35. Extremely low frequency (ELF) fields. Geneva: World Health Organization; 1984.

United Nations Environment Programme/World Health Organization/International Radiation Protection Association. Environmental health criteria 69. Magnetic fields. Geneva: World Health Organization; 1987.

Wertheimer, N.; Leeper, E. Electrical wiring configurations and childhood cancer. Am. J. Epidemiol. 109:273–284; 1979.

Zaffanella, L. E.; Deno, D. W. Electrostatic and electromagnetic effects of ultra-high-voltage transmission lines. Palo Alto, CA: Electric Power Research Institute; Final report, EPRI EL-802; 1978.

CHAPTER 7

INTERIM GUIDELINES ON LIMITS OF HUMAN EXPOSURE TO AIRBORNE ULTRASOUND

THE FIRST draft of these guidelines was completed by the International Non-Ionizing Radiation Committee (INIRC) of the International Radiation Protection Association (IRPA) at a meeting held in Rockville, Maryland in November 1981. Following approval by the IRPA Executive Council, the draft was distributed to Member Societies of IRPA, and to various institutions and individual scientists for comments. Many helpful comments and criticisms were obtained, and are gratefully acknowledged. Taking these comments into account, the IRPA/INIRC amended the guidelines and expanded the rationale.

The committee recognizes that when standards on exposure limits (ELs) are established, various value judgments are made. The validity of scientific reports has to be considered, and extrapolations from animal experiments to effects on humans have to be made. Cost vs. benefit analyses are necessary, including the economic impact of controls. The limits in these guidelines were based on the scientific data and no consideration was given to economic impact or other nonscientific priorities. However, from presently available knowledge, the limits should provide a safe, healthy working or living environment from exposure to airborne ultrasound under all normal conditions.

During the preparation of this document, the composition of the IRPA/INIRC was as follows:

H. P. Jammet, Chairman (France)
B. F. M. Bosnjakovic (Netherlands)
P. Czerski (Poland)
M. Faber (Denmark)
D. Harder (Germany)
J. Marshall (Great Britain)
M. H. Repacholi (Australia)
D. H. Sliney (U.S.A.)
J. C. Villforth (U.S.A.)
A. S. Duchêne, Scientific Secretary (France)

These guidelines were approved by the IRPA Executive Council on July 8, 1983, and published in *Health Physics* in April 1984.

INTRODUCTION

Ultrasonic energy is used in a wide variety of industrial processes, including cleaning, drilling, mixing, and emulsification. Most of these processes invariably emit airborne acoustic energy, not only at the ultrasonic operating frequency but also at sub-harmonics that, in many cases, are audible. Many industrial applications use high ultrasonic intensities that produce cavitation, observed as a type of boiling action in the liquid, which produces high audible noise levels.

In the industrial environment, many workers have complained of subjective symptoms (nausea, tinnitus, headaches, fatigue, etc.) when operating devices such as ultrasound cleaning tanks. Some data indicate hearing loss from exposure to very high intensities of airborne ultrasound, but no well-defined threshold for this effect has been determined.

Effects on the general public of airborne acoustic energy appear to be mediated by nervous reaction. Many people are unable to enter commercial establishments having an intrusion alarm (where the alarm is turned off, but the airborne ultrasound is still radiating) because they immediately suffer headaches or feel nauseated.

There are increasing numbers and varieties of consumer devices that use airborne ultrasound, including door openers, remote controls, intrusion alarms, pest repellers, and guidance devices for blind people. In general, these applications employ low intensity ultrasound in the frequency range 20 kHz–100 kHz. Many of these devices also have application in industry and commerce, and most operate predominantly at frequencies below 50 kHz.

The IRPA/INIRC in conjunction with the Division of Environmental Health, World Health Organization, Geneva, drafted a document entitled *Environmental Health Criteria for Ultrasound* (UN82). This document forms the primary scientific data base for the development of the following ELs for airborne ultrasound.

PURPOSE AND SCOPE

These guidelines are primarily aimed at protection against exposure from devices emitting high frequency airborne acoustic energy. Advice is provided and guidance given on limits of human exposure to airborne acoustic energy in the ultrasound range. Adverse health effects of exposure to airborne acoustic energy at levels normally encountered have been reported only at frequencies below 100 kHz and nearly all below 50 kHz. Thus, this standard has been limited to frequencies having one third-octave bands with mid frequencies from 20 kHz to 100 kHz.

BASIC CONCEPTS

Airborne ultrasound is usually quantified in terms of sound pressure level (SPL) in decibels (dB), such that:

$$\text{SPL (dB)} = 20 \log_{10}(p/p_r),$$

where p is the root mean square acoustic pressure and p_r is the reference pressure. p_r is equivalent to approximately the lowest level of audible sound perceived by humans at the most sensitive frequency (approx. 1 kHz), and is normally taken as equal to 20 micropascals (μPa). 20 μPa is equivalent to an acoustic intensity $I_r = 10^{-12}\text{W/m}^2$ in the air.

Since the acoustic intensity (I) is proportional to the square of the acoustic pressure, the sound pressure level can be equivalently expressed by

$$\text{SPL (dB)} = 10 \log_{10}(I/I_r).$$

Therefore, doubling the intensity (I) increases the SPL by 3 dB, whereas doubling the pressure (p) increases the SPL by 6 dB.

It should be noted that commonly used sound level meters have an effectively complete cut-off above about 20 kHz. Thus special sound level meters are needed for measurements above this frequency.

An octave band contains a range of frequencies, the upper frequency limit being twice the lower frequency limit. The centre or mid frequency used to designate each octave is twice the centre frequency of the preceding octave band. In this document, one-third octave bands are used to geometrically split an octave band into three parts and the mid frequency is used to designate each band.

EXPOSURE LIMITS

Limits of exposure to airborne ultrasound for occupational exposure are given in Table 1. The limits apply to continuous exposure to workers for an 8-hour working day. The limits in Table 1 may be increased in accordance with corrections given in Table 2, provided the total duration of exposure per day does not exceed 4 hours.

Exposure limits to airborne ultrasound for the general public are given in Table 3. The limits apply to continuous exposure to the general public for up to 24 hours per day.

National or local noise regulations or standards should incorporate limits of occupational and general public exposure to the auditory component of the airborne acoustic energy emitted from ultrasound devices.

Measurement of the sound pressure levels to determine adherence to the guidelines should be made at the normal height of the ears of exposed persons.

TABLE 1. *Limits for continuous occupational exposure to airborne ultrasound*

Mid frequency of one-third octave band (kHz)	Sound pressure level (dB re: 20 μ Pa)
20	75
25	110
31.5	110
40	110
50	110
63	110
80	110
100	110

TABLE 2. *Modification to occupational ELs given in Table 1 for exposure durations not exceeding 4 hours per day*

Total exposure duration (h)	Correction to SPL (dB)
2-4	+3
1-2	+6
0-1	+9

TABLE 3. *Limits for continuous exposure of the general public to airborne ultrasound*

Mid frequency of one third-octave band (kHz)	Sound pressure level (dB re: 20 μ Pa)
20	70
25	100
31.5	100
40	100
50	100
63	100
80	100
100	100

A brief rationale for the ELs is given in Appendix 1. Measures needed to ensure compliance with these limits are given in Appendix 2.

EXCLUSIONS

The limits recommended in these guidelines may be exceeded for occupational exposure, if workers are provided with ear protectors that reduce the ultrasound levels at their ears to the sound pressure levels given in Table 1. No exclusions to these limits are recommended for exposure of the general public (Table 3).

CONCLUDING REMARKS

Since the data on the effects of human exposure to airborne ultrasound are presently limited, these guidelines will be subject to periodic revisions and amendments with advance of knowledge.

REFERENCES

Ac67 Acton W. I. and Carson M. B., 1967, "Auditory and subjective effects of airborne noise from industrial ultrasonic sources", *Br. J. Ind. Med.* **24,** 297–304.

Ac74 Acton W. I., 1974, "The effects of industrial airborne ultrasound on humans", *Ultrasonics* **12,** 124–128.

Ac75 Acton W. I., 1975, "Exposure criteria for industrial ultrasound", *Ann. Occup. Hyg.* **18,** 267–268.

Ac77 Acton W. I. and Hill C. R., 1977, "Hazards of industrial ultrasound", *Protection* **14**(19), 12–17.

Cr77 Crabtree R. B. and Forshaw S. E., 1977 *Exposure to Ultrasonic Cleaner Noise in the Canadian Forces,* Dept. of National Defence, Ottawa, DCIEM Technical Report No. 77 × 45.

He81 Herman B. A. and Powell D., 1981, *Airborne Ultrasound: Measurement and Possible Adverse Effects,* U.S. Dept. of Health and Human Services, Rockville, MD, Pub. (FDA) 81–8163.

Hi82 Hill C. R. and ter Haar G., 1982, "Ultrasound", in: *Non-Ionizing Radiation Protection* (Edited by M. J. Suess), WHO Regional Publications, European Series No. 10, World Health Organization, Regional Office for Europe, Copenhagen, Denmark, pp. 199–228.

HW79 Health and Welfare Canada, 1979, *Noise Hazard and Control,* Health and Welfare Canada, Ottawa, Pub. 79-EHD-29.

HW80 Health and Welfare Canada, 1980, *Guidelines for the Safe Use of Ultrasound, Part II, Industrial and Commercial Applications,* Safety Code-24, Health and Welfare Canada, Ottawa, Pub. 80-EHD-60.

IL77 International Labour Office, 1977, *Protection of Workers Against Noise and Vibration in the Working Environment.* ILO Codes of Practice, Geneva, Pub. No. ISBN 92-2-101709.

Re81 Repacholi M. H., 1981, *Ultrasound: Characteristics and Biological Action,* National Research Council of Canada, Ottawa, Canada, K1A OR6, Pub. No. 19244.

Sk65 Skillern C. P., 1965, "Human response to measured sound pressure levels from ultrasonic devices", *Am. Ind. Hygiene J.* **26,** 132–136.

UN80 United Nations Environment Programme/World Health Organization 1980, *Noise,* Environmental Health Criteria No. 12, WHO, Geneva.

UN82 United Nations Environment Programme/World Health Organization/International Radiation Protection Association, 1982, *Ultrasound,* Environmental Health Criteria No. 22, WHO, Geneva.

APPENDIX 1: RATIONALE FOR EXPOSURE LIMITS

The biological effects of exposure to airborne ultrasound have recently received a thorough review (UN82). They have previously been reviewed by various authors and organizations (Ac67; Ac74; Ac77; He81; Hi82; HW80; Re81).

Much less than 1% of the airborne ultrasound is absorbed by human skin, the rest is reflected. Hair strongly absorbs sound and ultrasound in the frequency range of interest. The ear, being a more efficient coupler of airborne acoustic energy than any other part of the human body, is considered the most sensitive organ.

Acton and Carson (Ac67) failed to detect either temporary or permanent losses of hearing in industrial workers exposed to levels of airborne ultrasound of approximately 120 dB. On the other hand, temporary threshold shifts were detected in hearing of subjects exposed to frequencies of 18 kHz at 150 dB for approximately 5 minutes (Ac67).

The more sensitive indicator of potential harm from airborne ultrasound exposure comes from reports of subjective effects—nausea, headaches, fatigue, tinnitus, or an unpleasant sensation of fullness or pressure in the ears. Workers with good hearing at the upper frequencies of the audible range were reported to complain of these subjective effects (Ac67; Ac74; Ac77; Cr77; He81; Hi82; Re81; Sk65). It was suggested that these effects were produced by audible components of the ultrasonic frequency (Ac67; Ac74; Ac77). Airborne sound at levels of approximately 78 dB at 16 kHz was reported to cause subjective symptoms, while levels of 100 dB at 20 kHz and 25 kHz did not cause these effects. However, Crabtree and Forshaw (Cr77) reported subjective symptoms in Canadian Forces personnel working around ultrasonic cleaning tanks. The SPL in the 20 kHz one-third octave band did not exceed 105 dB at the operator's position.

A recommended occupational EL of 110 dB for frequencies with mid frequencies of one-third octave bands above 20 kHz seems well justified from the available data (UN82). What seems to differ in many standards is the EL at 20 kHz, mid frequency of one-third octave band (UN82). Presently available data do not provide a threshold for effects in this frequency band. Acton (Ac75) recommends an SPL of 75 dB for the one-third octave band with mid frequency of 20 kHz. His rationale was that the nominal frequency limits of the one-third octave band centered on 20 kHz are 17.6 kHz to 22.5 kHz and the lower end is within the upper end of the audible frequency range of many people who operate industrial ultrasonic devices. The SPL of 75 dB seems appropriate from presently available data. Sound pressure levels of 100 dB at frequencies in the range 17.6 kHz to 20 kHz may produce severe auditory and subjective effects, although permanent hearing loss is unlikely (Ac67).

When considering ELs for the general public, a number of additional factors must be taken into account.

1. Exposure may occur for up to 24 hours per day.
2. There is no medical surveillance as is possible for a controlled occupational group.
3. It would be undesirable to require hearing protectors or other protective devices to keep levels at the ears within the limits.
4. Noise-related effects such as annoyance, stress, etc., must be considered in addition to other possible auditory effects.
5. The general public is a population containing a broad range of sensitivities to insult from physical agents.

Existing data suggest that exposure of the general public to airborne ultrasound (one-third octave band mid frequencies above 20 kHz) at levels up to 110 dB is not known to cause untoward health effects. However, noting that the general population can potentially be exposed 24 hours per day and for the other considerations noted above, an added safety factor should be incorporated, at least as an interim measure until more definite data on adverse health effects of exposure to airborne ultrasound become available. Thus an SPL of 100 dB is recommended. For similar reasons, an added safety factor should be incorporated into the EL for frequencies in the range of the one-third octave band centered on 20 kHz. An SPL of 70 dB is recommended, although it is noted that noise ELs in each country (HW79; IL77; UN80) may cover audible frequencies up to 20 kHz.

APPENDIX 2: PROTECTIVE MEASURES

Adoption of ELs constitutes the first step in protection. Equipment performance and emission standards should then be derived from the ELs.

Standardized measurement techniques and survey procedures should be introduced and adhered to. Determination of sound pressure levels at given locations in an airborne acoustic field is normally made using a device consisting of a capacitor microphone having a flat response over the frequency range of interest, associated electronics, and a set of one-third octave filters. The audible component of an acoustic field can be measured using a sound level meter with a low-pass filter to reject frequencies above 20 kHz.

In air, ultrasound at a frequency of 40 kHz has a wavelength of about 8.5 mm, is quite directional (depending on the ratio of the source diameter and the wavelength), is easily attenuated by barriers, and loses about 0.06 dB for every 30 cm the wave travels due to absorption by the air. At lower frequencies, lower attenuation occurs due to air absorption and vice versa. In addition, the acoustic intensity radiating from a point source reduces 6 dB for each doubling of distance in the far field. The airborne acoustic field can be extremely complex and careful mapping of the field should be made during surveys since levels can vary significantly over a short distance. Preliminary measurements should be made to assess the acoustic field present. As most acoustic sources vary in intensity level and frequency spectrum, careful consideration of each measurement situation is essential to obtain meaningful data.

Measurement of SPLs should be made with the microphone positioned at the ear height of exposed persons where possible. For making these measurements, it is highly desirable to use a small diameter microphone with adequate response at high frequencies and good directional properties.

Microphones and filters can have frequency-dependent errors, and so must be calibrated before use. Complete calibration of the measurement equipment is complex and should be conducted by a qualified laboratory or the equipment manufacturer. Before and during measurements in the field, checks on the accuracy of the equipment should be made. Such checks are normally made using a field calibrator, which generates a known acoustic signal. The highest frequency of a field calibrator is normally limited to 2 kHz. However, such a calibrator should still be used because the equipment normally performs correctly either at all frequencies or at none. Thus the calibrator will allow the detection of malfunctions related to frequency. The accuracy of measurement instruments should be within ±2 dB for frequencies up to 25 kHz and ±5 dB above 25 kHz to the upper frequency limit of the instrument (HW80).

Responsibility for protection and safety should be clearly assigned to appropriate regulatory agencies, health departments, and individuals. Where possible and needed, principles of medical and environmental surveillance should be developed and appropriate responsibilities assigned (HW80; IL77).

Responsibility for the evaluation of protection from emissions from newly designed equipment and installations should be assigned and implemented. This should also allow for the development of safe use guidelines or codes of practice. These should constitute an integral part of instructions for the user and for the development of an emission standard (where applicable), before such equipment is mass-produced. One might consider pre-market approval of these devices on the basis of safety. Wherever possible, exposure controls should be built into mass-produced equipment to avoid dependence for safety on instructions and labels alone.

Guidelines on measures for protecting humans against exposure to acoustic energy have recently been published (HW80; IL77). Both these documents are good references for persons or institutions developing their own codes of safe practice.

CHAPTER 8

ALLEGED RADIATION RISKS FROM VISUAL DISPLAY UNITS—A STATEMENT

VISUAL display units (VDUs) have become a major element in the modern work environment as an interface between man and computer. The discussion as to whether work at VDUs can affect human health has been centered on different types of effects, such as eye damage or discomforts, neck and shoulder discomfort, different stress reactions, skin disorders, and adverse reproductive outcomes. In this context, much concern has been expressed in the media in relation to the possibility of radiation hazards due to VDUs based on cathode ray tubes (CRTs). This aspect will be covered in the present document, while the reader is referred to other texts for a discussion on the influence of various ergonomic factors on health (see, e.g., World Health Organization, 1987).

The present statement was approved by the International Non-Ionizing Radiation Committee of the International Radiation Protection Association (IRPA/INIRC) in March 1987. During its preparation, the composition of the IRPA/INIRC was as follows:

H. P. Jammet, Chairman (France)
J. Bernhardt (Federal Republic of Germany)
B. Bosnjakovic (Netherlands)
P. Czerski (U.S.A.)
M. Grandolfo (Italy)
D. Harder (Federal Republic of Germany)
B. Knave (Sweden)
J. Marshall (Great Britain)
M. Repacholi (Australia)
D. H. Sliney (U.S.A.)
J. Stolwijk (U.S.A.)
A. S. Duchêne, Scientific Secretary (France)

A number of careful scientific studies have been focused on the measurement of electromagnetic radiation and fields due to VDUs, while some limited attention has also been given to acoustic radiation; several publications also address the topic of health risk assessment (see reference list).

1. Soft x-ray radiation is produced within the CRT. The glass material of the tube, however, effectively prevents any leakage of x-ray radiation outside of the tube during operation. Thus, x-ray radiation from VDUs is not detectable with normal measuring instruments.

2. Ultraviolet radiation in the near region (UV-A) can be detected from certain VDUs. The levels are, however, insignificant compared with present occupational standards (see Chapter 3) and also insignificant compared with emission from other sources (e.g., sunlight through windows). In one investigation, VDU operators were found to be exposed to lower levels of UV-A than those not working with VDUs, attributable to the fact that the former often draw the window curtains.

3. Visible radiation can be measured and is necessary in order to perform the intended function of the CRT—to provide a visual display. Luminance levels recorded are far below current exposure limits, thus precluding (according to present knowledge) the possibility

of pathological injury due to excessive exposure. There are ergonomic considerations of light emission from the display, such as flicker, contrast glare, or readability. These are, however, not considered in this context.

4. Infrared (IR) radiation is emitted from all warm bodies, and since all surfaces of the VDU are at room temperature or slightly above, IR radiation can be detected, although at levels far below any levels of concern for potential hazards.

5. In the extremely low frequency and the radiofrequency regions, electric and magnetic fields have been detected. The dominant sources are the power supply and the vertical and horizontal sweep arrangements (at frequencies of some 50–80 Hz and 15–35 kHz, respectively). Compared with fields in many industrial or household situations, the fields around VDUs do not correspond to high exposure situations. These fields do not appear to represent any risk factor when evaluated by comparison with current standards, guidelines, and recommendations for occupational exposure.

6. In some countries, a number of VDU operators have experienced skin disorders. The relationship of these to VDU work is not known. Electrostatic fields at VDU workplaces have been suggested as a possible cause of skin disorders. Research conducted hitherto has indicated that the electrostatic charge of the operator might be a relevant factor. A relationship between electrostatic fields and skin disorders must, however, still be regarded as hypothetical.

7. Airborne ultrasonic (acoustical) radiation is produced in VDUs as a result of mechanical vibrations generated in the core of the flyback transformer (responsible for the horizontal sweep of some 15–35 kHz). The sound pressure levels found are considerably below existing exposure limits (75 dB) (see Chapter 7). Some sensitive individuals may detect this sound or a subharmonic as an annoying factor.

Effects that have been suggested as caused by exposure to electromagnetic radiation or fields include adverse pregnancy outcome or cataracts. Comparison of the occurrence of cataracts or of adverse pregnancy outcome among VDU operators with those of controls have failed to show an excess occurrence due to VDU work.

In conclusion, based on current biomedical knowledge, there are no health hazards associated with radiation or fields from VDUs. Thus, there is no scientific basis to justify shielding or radiation monitoring of VDUs. However, since a large number of people are involved in VDU work, it is important that further knowledge is attained on certain areas where our knowledge must be regarded as incomplete. These areas include: (a) further investigations into the possibility that skin disorders may be related to VDU work, and if so, the factor(s) involved; and (b) the possibility of interactions between low frequency magnetic fields and biological systems. Considerations should be given to magnetic fields in various situations and should not be restricted to VDU work situations.

Measures should be taken to ensure that VDU work places are ergonomically well designed. This includes aspects of the VDU, the work station, and work environment, as well as work organization. Visual screening examination is also valuable, in ensuring that the operator has adequate visual acuity, and that any corrective glasses are suitable for use at the VDU working distance.

BIBLIOGRAPHY

Bergqvist, U. O. V. Video display terminals and health. Scand. J. Work. Environ. Health. 10(suppl. 2):1–87; 1984.

Bureau of Radiological Health. An evaluation of radiation emission from video display terminals. Rockville, MD: U.S. Food and Drug Administration; HHS Publication FDA 81-8153; 1981.

Cox, E. A. Radiation emission from visual display units. In: Pearce, B. G., ed. Health hazards of VDTs? Chichester, Great Britain: John Wiley; 1984: 25–37.

Guy, A. W. Health hazards assessment of radio frequency electromagnetic fields emitted by video display terminals. Report prepared for IBM, Office of the Director of Health and Safety, Corporate Headquarters, Armonk, NY; 2 December 1984.

Harvey, S. M. Electric-field exposure of persons using video display units. Bioelectromagnetics 5:1–12; 1984.

Institut de recherche en santé et en sécurité du travail du Québec. Report of the task force on video display terminals and workers' health. Montréal: IRSSTQ; S-003; May 1984.

Marriott, I. A.; Stuchly, M. A. Health aspects of work with video display terminals. J. Occup. Med. 28:833–848; 1986.

National Board of Occupational Safety and Health. Electromagnetic radiation and fields at visual display terminals (VDTs). Solna, Sweden: NBOSH; 1986.

Paulsson, L. E.; Kristiansson, I.; Malmström, I. Strålning från dataskärmar (Radiation from VDTs). Stockholm, Sweden: Statens Strålskyddsinstitut; a 84–08; 1984.

Pomroy, C.; Noel, L. Low-background radiation measurements on video display terminals. Health Phys. 46:413–417; 1984.

Repacholi, M. H. Video display terminals—should operators be concerned? Australasian Phys. Eng. Sci. Med. 8(2):51–61; 1985.

Stuchly, M. A.; Lecuyer, D. W.; Mann, R. D. Extremely low frequency electromagnetic emissions from video display terminals and other devices. Health Phys. 45:713–722; 1983.

Stuchly, M. A.; Repacholi, M. H.; Lecuyer, D. W.; Mann, R. D. Radiofrequency emissions from video display terminals. Health Phys. 45:772–775; 1983.

Wolbarsht, M. L.; O'Foghludha, F. A.; Sliney, D. H.; Guy, A. W.; Smith, A. A.; Johnson, G. A. Electromagnetic emission from visual display units. A non-hazard. In: Non-ionizing radiation (proc. of a topical symp.). Cincinnati, OH: American Conference of Governmental Industrial Hygienists; 1980;193–200.

World Health Organization. Visual display terminals and workers health. Geneva: WHO; offset publication no. 99; 1987.

Zuk, W. M.; Stuchly, M. A.; Dvorak, P.; Deslauriers, Y. Investigation of radiation emissions from video display terminals. Ottawa: Health and Welfare Canada, Radiation Protection Bureau; 83-EHD-91; 1983.

CHAPTER 9

FLUORESCENT LIGHTING AND MALIGNANT MELANOMA

DURING the last 20 years, epidemiological studies of the incidence of malignant melanoma of the skin have suggested an etiologic role of ultraviolet radiation (UVR) (Anaise et al., 1978; Beral and Robinson, 1981; Dubin et al., 1986; Fears et al., 1976; Gallagher et al., 1987; Green et al., 1986; Holman et al., 1986; Klepp and Magnus, 1979; Kripke, 1979; Lee, 1985; MacKie, 1987; Sober, 1987).

More recently, some concern has arisen regarding the potential role that UVR emitted from fluorescent lighting could play in the etiology of malignant melanoma. This concern developed from published reports of epidemiological studies of the incidence of malignant melanoma in various occupations. An initial study, conducted in Australia by Beral et al. (1982), indicated that indoor office workers who reported working for many years under fluorescent lights had a higher incidence of malignant melanoma than those who reported they did not and those who worked outdoors. Further epidemiological studies specifically designed to test the hypothesis that exposure to UVR from artificial sources correlated with malignant melanoma incidence have shown either a positive correlation (Elwood, 1986; English et al., 1985; Pasternack et al., 1983) or no significant correlation (Pasternack et al., 1983; Rigel et al., 1983; Sorahan and Grimley, 1985; Swerdlow et al., 1988). The results of these studies and problems in their methodologies have been reviewed (Muel et al., 1988).

Despite the fact that Beral's original reported correlation was greatest for malignant melanoma occurrences on the trunk, which is normally covered by clothes, it has been speculated that the weak UVR emissions of fluorescent lamps could explain the correlation. This hypothesis would require an indirect mechanism, such as a general UVR effect upon the immune system (Kripke, 1979).

An obvious difficulty in all of the above studies has been the estimation of indoor vs. outdoor exposure to UVR and reliance on the recollections of both patients and matched controls of the characteristics of indoor lighting. In an ongoing study in Queensland, Australia (private communication [1988]. R. MacLennan, Queensland, Australia), poor correlation was found between the recollections of patients and controls about the types of artificial lighting to which they had been exposed and the descriptions provided by a lighting engineer familiar with the buildings where the patients and controls had worked. It is therefore likely that one cannot completely rely upon the results found in such studies. From a theoretical point of view, it is far more likely that alternative theories are more valid to explain the clear increase in incidence of malignant melanoma of the skin among these indoor workers.

Recreational and social habits of indoor workers may increase the likelihood that they would sustain severe, acute overexposure to sunlight on untanned, normally covered regions of the skin (Carlton-Foss, 1982; Holman et al., 1986; MacKie and Aitchison, 1982). The very low incidence of melanoma lesions on least-exposed areas of the skin (e.g., buttocks and women's breasts) and the high incidence of appearance on normally covered areas of the body (e.g., the trunk) argues for the hypothesis of an etiology related to acute overexposure to sunlight and against the hypothesis of an etiology related to direct fluorescent light exposure. The role of UV-B in an indirect photochemical mechanism remains unresolved (Wiskemann et al., 1986).

The very low contribution of UV-B (280–315 nm) and UV-C (100–280 nm) radiation (CIE, 1983; Cole et al., 1985; Diethelm, 1970; Muel et al., 1988; Sliney and Wolbarsht, 1980; Whillock et al., 1988) to the emissions from fluorescent lighting also argues against this exposure as a major or very significant etiology. Plastic diffusers serve as shields and will further attenuate UV-B and UV-C emissions (Cole et al., 1985; Whillock et al., 1988). Even if individuals were exposed to the UVR from unshielded fluorescent lamps, the exposure from these would be only a small fraction of exposure to solar UVR (Muel et al., 1988).

At least one health authority has suggested that where people are concerned about the UVR emitted by fluorescent lamps, absorbing plastic diffusers may be fitted to effectively remove it (FDA, 1988). However, the reported health benefits of low-level exposure to UV-B (e.g., vitamin D production, adaptation to further exposure) have been used to argue in favor of people regularly receiving a small amount of UV-B (Health Council of the Netherlands, 1986).

Considering all of the above arguments, the International Non-Ionizing Radiation Committee of the International Radiation Protection Association (IRPA/INIRC) concludes that UVR exposure from indoor fluorescent lighting should not be considered a malignant melanoma risk. Lamp manufacturers should be aware of this issue and not allow the current low levels of emission of UV-B from fluorescent lighting to increase.

The present statement was approved by the IRPA/INIRC during its meeting in May 1989.

REFERENCES

Anaise, D. R.; Steinitz, R.; Hur, N. B. Solar radiation: A possible etiological factor in Israel. A retrospective study (1960–1972). Cancer 42(1):299–304; 1978.

Beral, V.; Evans, S.; Shaw, H.; Milton, G. Malignant melanoma and exposure to fluorescent lighting at work. The Lancet II: 290–293; 1982.

Beral, V.; Robinson, N. The relationship of malignant melanoma, basal and squamous skin cancers to indoor and outdoor work. Brit. J. Cancer 44:886; 1981.

Carlton-Foss, J. A. Letter: Comments on Beral's paper. The Lancet II:818; 1982.

CIE, Centre d'information de l'éclairage. Fluorescent lamps and health. Jan. 1983 Hors Série No. 8; 1983. Available from: CIE, 52 Boulevard Malesherbes, 75008 Paris, France.

Cole, C.; Forbes, P. D.; Davies, R. E.; Urbach, F. Effect of indoor lighting on normal skin. Ann. NY Acad. Sci. 453:305–316; 1985.

Diethelm, R. Can fluorescent tubes cause skin carcinomas? Schweizerische Medizinische Wochenschrift—Journal Suisse de Médecine 100: 1159–1160; 1970.

Dubin, N.; Moseson, M.; Pasternack, B. S. Epidemiology of malignant melanoma, pigmentary traits, ultraviolet radiation. In: Gallagher, R. P., ed. Epidemiology of malignant melanoma. Recent results in cancer research: 102. Heidelberg: Springer-Verlag; 1986:56–75.

Elwood, J. M. Could melanoma be caused by fluorescent light? A review of relevant epidemiology. In: Gallagher, R. P., ed. Epidemiology of malignant melanoma. Recent results in cancer research: 102. Heidelberg: Springer-Verlag; 1986: 127–136.

English, D. R.; Rouse, I. L.; Xu, Z.; Watt, J. D.; Holman, C. D. J.; Heenan, P. J.; Armstrong, B. K. Cutaneous malignant melanoma and fluorescent lighting. J. Nat. Cancer Inst. 74:1191–1197; 1985.

Food and Drug Administration, Center for Devices and Radiological Health. Fact Sheet 'Fluorescent Lamps.' Rockville, MD: U.S. Food and Drug Administration; June 1988.

Fears, I. R.; Scotto, J.; Schneiderman, M. A. Skin cancer, melanoma and sunlight. Am. J. Public Health 66:461–464; 1976.

Gallagher, R. P.; Elwood, J. M.; Threlfall, W. J.; Spinelli, J. J.; Fincham, S.; Hill, G. B. Socioeconomic status, sunlight exposure, and risk of malignant melanoma: The Western Canada Melanoma Study. J. Nat. Cancer Inst. 79:647–652; 1987.

Green, A.; Bain, C.; MacLennan, R.; Siskind, V. Risk factors for cutaneous melanoma in Queensland. In: Gallagher, R. P., ed. Epidemiology of malignant melanoma. Recent results in cancer research: 102. Heidelberg: Springer-Verlag; 1986: 76–97.

Health Council of The Netherlands. Human exposure to UV radiation. Report 1986/9, Gezondheidsraad, PO Box 90517, 2509 LM, The Hague; 1986.

Holman, C. D. J.; Armstrong, B. K.; Heenan, P. J. Relationship of cutaneous malignant melanoma to individual sunlight exposure habits. J. Nat. Cancer Inst. 76:403–414; 1986.

Klepp, O.; Magnus, K. Some environmental and bodily characteristics of melanoma patients. A case-control study. Int. J. Cancer 23:482–486; 1979.

Kripke, M. L. Speculations on the role of ultraviolet radiation in the development of malignant melanoma. J. Nat. Cancer Inst. 65:541–545; 1979.

Lee, J. A. H. The rising incidence of cutaneous malignant melanoma in Scotland. Am. J. Dermatopathology 7(Suppl.):35–39; 1985.

MacKie, R. M. Links between exposure to ultraviolet radiation and skin cancer. A report of the Royal College of Physicians. J. Roy. Coll. Physicians 21:91–96; 1987.

MacKie, R. M.; Aitchison, T. Severe sunburn and subsequent risk of primary cutaneous malignant melanoma in Scotland. Br. J. Cancer 46:955–960; 1982.

Muel, B.; Cesarini, J. P.; Elwood, J. M. Malignant melanoma and fluorescent lighting. C.I.E. J. 7(1): 29–33; 1988.

Pasternack, B. S.; Dubin, N.; Moseson, M. Letter: Malignant melanoma and exposure to fluorescent lighting. The Lancet I:704; 1983.

Rigel, D. S.; Friedman, R. J.; Levenstein, M.; Greenwald, D. I. Relationship of fluorescent lights to malignant melanoma: Another view. J. Der. Sur. Oncol. 9:836–838; 1983.

Sliney, D.; Wolbarsht, M. Safety with lasers and other optical sources. New York: Plenum Press; 1980.

Sober, A. J. Solar exposure in the etiology of cutaneous melanoma. Photodermatology 4:23–31; 1987.

Sorahan, T.; Grimley, R. P. The etiological significance of sunlight and fluorescent lighting in malignant melanoma: A case-control study. Br. J. Cancer 52:765–769; 1985.

Swerdlow, A. J.; English, J. S. C.; MacKie, R. M.; O'Doherty, C. J.; Clark, J.; Hole, D. J. Fluorescent lights, ultraviolet lamps and risk of cutaneous melanoma. Brit. Med. J. 297:647–650; 1988.

Whillock, M. J.; Clark, I. E.; McKinlay, A. F.; Todd, C. D.; Mundy, S. J. Ultraviolet levels associated with the use of fluorescent general lighting, UVA and UVB lamps in the workplace and home. Oxon: National Radiological Protection Board; NRPB R221; 1988.

Wiskemann, A.; Sturm, E.; Klehr, N. W. Fluorescent lighting enhances chemically induced papilloma formation and increases susceptibility to tumour challenge in mice. J. Cancer Res. Clinical Oncology 112:141–143; 1986.

INDEX